FOUR CORNERS

Also by Kenneth A. Brown

Cycles of Rock and Water on the Pacific Edge
Inventors at Work

FOUR CORNERS

HISTORY, LAND, AND PEOPLE OF THE DESERT SOUTHWEST

Kenneth A. Brown

HarperPerennial
A Division of HarperCollins*Publishers*

A hardcover edition of this book was published in 1995 by HarperCollins Publishers.

FOUR CORNERS. Copyright © 1995 by Kenneth A. Brown. All rights reserved. Printed in the United States of America. No part of this book may be used or reproduced in any manner whatsoever without written permission except in the case of brief quotations embodied in critical articles and reviews. For information, address HarperCollins Publishers, Inc., 10 East 53rd Street, New York, NY 10022.

HarperCollins books may be purchased for educational, business, or sales promotional use. For information please write: Special Markets Department, HarperCollins Publishers, Inc., 10 East 53rd Street, New York, NY 10022.

First HarperPerennial edition published in 1996.

Designed by Laura Lindgren
Map by Paul Pugliese

The Library of Congress has catalogued the hardcover edition as follows:

Brown, Kenneth A.
    Four corners : history, land, and people of the desert South-
west / Kenneth A. Brown. — 1st ed.
      p.   cm.
    Includes index.
    ISBN 0-06-016756-4
    1. Human geography—Four Corners Region.   2. Geology—Four
Corners Region.   3. Natural history—Four Corners Region.   4. Indians
of North America—Four Corners Region—History.   5. Four Corners
Region—History.   6. Four Corners Region—Description and travel.
I. Title.
GF504.F68B76   1995
304.2'09789—dc20   95-40208

ISBN 0-06-092759-3 (pbk.)
96 97 98 99 00 ❖ / RRD 10 9 8 7 6 5 4 3 2 1

For Marion and Julia

CONTENTS

Photographs follow page 148.

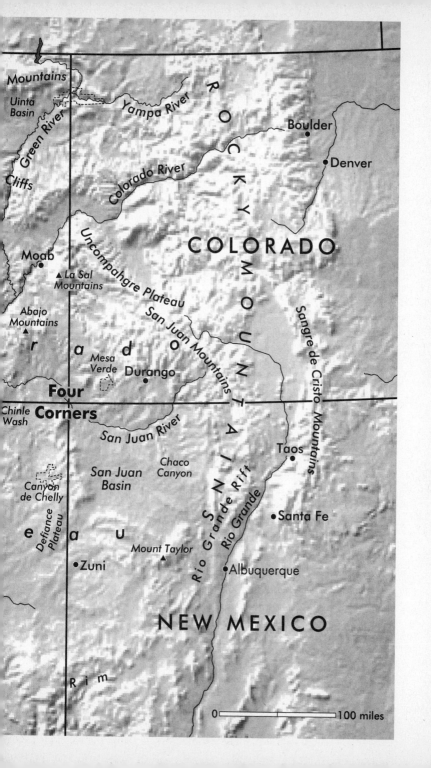

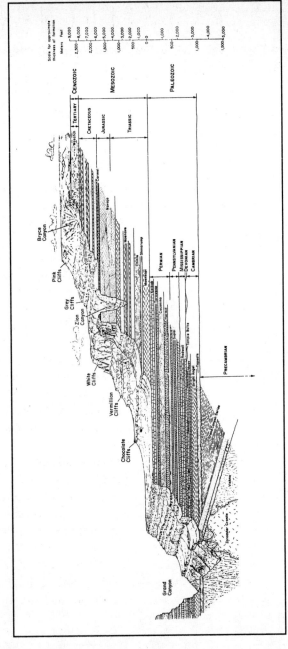

W. K. HAMBLIN

ACKNOWLEDGMENTS

I SPENT JUST over a year in the Southwest piecing together the travels and interviews that make up this book. Many of those I met and talked to are included in these pages, but others are not, and I would like to acknowledge their help and assistance.

For my work on the archaeology of the Southwest I would like to thank Terry Coriell of Mancos, Colorado, who was a constant source of help and information, introducing me to archaeologists in Colorado and New Mexico and taking me on trips to see pueblo ruins and rock art throughout the region. Others like Karen Adams of Crow Canyon Archaeological Center in Cortez, Colorado, and Blaine Phillips with the Bureau of Land Management in Vernal, Utah, also gave generously of their time. I received a variety of help from the libraries and staffs of both the Crow Canyon Archaeological Center in Cortez, Colorado, and the Anasazi Heritage Center in Dolores, Colorado. Thanks too to La Plata Archaeology of Cortez, Colorado, for inviting me out to their field sites and excavations.

For my work on the biology of desert and high mountain regions of the area I would like to say a special thanks to David Tippetts with the U.S. Forest Service's Intermountain Research Center for arranging interviews and meetings with a wide variety of Forest Service Personnel, including staff and scientists at the agency's Shrub Sciences Laboratory in Provo, Utah. Steve Monson provided me with a valuable insight on the finer points of desert rangelands and the efforts currently underway to restore them. Thanks as well to the Utah Fish and Game Department and the Intermountain Research Station in Ephraim, Utah.

For my work on the geology of the Four Corners region I was helped by a number of scientists working in the region: Pete Peterson, Tom Fouch, Bob Scott, Peter Rowley, and Russell Dubiel with the U.S. Geological Survey in Denver, Colorado, and Rich Hereford with the U.S. Geological Survey in Flagstaff, Arizona.

My work in the Canyonlands and Arches National Park area was helped by Larry Fredericks with the National Park Service, who provided me with contacts and information on the parks. For my travels on the Navajo Reservation I would like to thank Anna Lou Pickett for her introduction to many friends and sources, including Gene and Daisy Thompson, Jenny Notah, and Richard Begay with the tribe's cultural preservation office. Thanks too to Angela Barney Nez, who helped arrange my visit with her father. For my time at Zuni I would like to thank Sister Rose Marie Deibel and the sisters, brothers, and staff at St. Anthony's Zuni Indian Mission, who provided me with not only an introduction to many families and people within the tribe, but a warm and friendly place to stay as well. Thanks also to Dolly and Albert Banteah and family of Zuni, who always made me welcome. Martin Link gave unstintingly of his time, sharing his ideas and perceptions of regional history with me in several lengthy interviews and conversations at his office in Gallup, New Mexico. Thanks too to George Otero of Taos, New Mexico, for his introduction to people and ideas in northern New Mexico.

Thanks to Pat and Julia La Chapelle of Mancos, Colorado; Keith Paris and Linda Delgado of Durango, Colorado; and the Arsenio Romero family of Cordova, New Mexico, who helped provide a home away from home while I was traveling.

Behind the scenes, a number of others helped bring this book to completion. Thanks to Larry Ashmead at HarperCollins for the ideas and support that started this book in motion and to my agent John Ware for his constant support and perspective. Much thanks are due as well to my editor Jason Kaufman at HarperCollins, whose energy and enthusiasm helped keep this book alive long after my own had faded. His careful editing and ideas helped bring the book to life. Thanks too to my parents for their tireless patience and support over the writing of this book.

FOUR CORNERS

BOOK ONE

Beginnings

IN CALLING UP images of the past, I find that the plains of Patagonia frequently cross before my eyes; yet these plains are pronounced by all that is all wretched and useless.

They can be described only by negative characters; without habitations, without water, without trees, without mountains, they support merely a few dwarf plants. Why then, and the case is not peculiar to myself, have these arid wastes taken so firm a hold on my memory? Why have not the still more level, the greener and more fertile Pampas, which are serviceable to mankind, produced an equal impression?

I can scarcely analyze these feelings: but it must be partly owing to the free scope given to the imagination.

The plains of Patagonia are boundless, for they are scarcely passable, and hence unknown; they bear the stamp of having lasted, as they are now, for ages, and there seems no limit to their duration through future time. If, as the ancients supposed, the flat earth was surrounded by an impassable breadth of water, or by deserts heated to an intolerable excess, who would not look to these last boundaries to man's knowledge with deep but ill-defined sensations?

CHARLES DARWIN
The Voyage of the Beagle, 1836

O N E

Island in the Sky

THE SPRING ON the canyon floor is fringed with reeds and tall grass. After a long dry spell it shrinks to little more than a patch of wet sand. Over the past few months, however, the rains have been plentiful and the spring now lies at the bottom of a broad, shallow pool. So much water is bubbling up out of the ground that the surface of the pool seems to shimmer and dance. At the far end it slides over a rounded ledge of sandstone, joining the narrow stream that winds through the floor of the canyon toward the Colorado River some five miles away.

Water is valuable in the desert, and the sand near the spring is covered with tracks: the footprints of coyotes and mule deer that roam the desert at night. At midday the canyon seems almost lifeless. The stillness is broken only by scattered, solitary sounds: the croak of a raven flying overhead, the brittle rustle of dry grass and old weeds, the clatter of loose stones kicked up by a desert bighorn sheep traversing a scree slope near the canyon rim some five hundred feet above.

It is late fall, and the air is still and dry. Down in the canyons among the walls of red and cream-colored rocks the leaves on the cottonwood trees are bright gold. Up above, the sky is the color of turquoise and laced with thin, white ribbonlike clouds. Nights are clear and cold, illuminated by the soft, blue light of the moon. Mornings I walk down from my camp on a sandy ledge in a nearby side canyon to find the spring rimmed with ice. In three days the high desert will be covered with more than five inches of snow.

I am somewhere on the edge of Canyonlands National Park, not far from the junction of the Colorado and Green rivers. I have not seen anyone for two weeks. Five miles overhead, planes bound for Los Angeles, Chicago, and Phoenix arc across the sky. Down on the ground the world is a wilderness of bare rock. The nearest paved road is more than twenty miles away as the raven flies; the nearest town more than forty. On foot, however, the distance to those familiar points of reference is considerably farther, broken by a seemingly endless maze of canyons and mesas. Here in the labyrinth of canyons that wind through the slickrock deserts of southeastern Utah, just reaching the rim can take several hours: squeezing up slots and cracks in the sheer, smooth walls of rock through a progression of ledges and handholds that leads steadily upward.

Up on the canyon rim you can lean against the twisted trunk of a juniper tree and look out across a surreal landscape of spires, fins, and arches of rock that stretches for miles in every direction. Here and there domes of smooth, worn rock rise up out of the ground like giant waves, colored in shades of red, pink, and white. The air is so sharp and clear that you can see the tops of high, snow-covered peaks more than seventy miles away.

The high desert is a landscape of sharp contrasts. Although the ground here is more than a mile high, it is not rich and green like the alpine reaches of the Rockies or the Sierras, but dry and bare. While summer temperatures frequently climb above a hundred degrees, in winter it can plunge to more than twenty below zero. Two months of dry, rainless heat can be broken by a late-summer thunderstorm that drops more than four inches of rain in an hour. Flash floods fill dry washes with five- and ten-foot-high surges of water that disappear almost as quickly as they come. Other storms drift across the landscape without leaving a drop: purple and black thunderheads twenty to thirty thousand feet high with silver and gray curtains of rain trailing beneath them that evaporate without even reaching the ground.

At first glance the land here seems almost uninhabitable. Mesa tops and canyon rims are covered by scrubby forests of pinyon and juniper. Elsewhere the ground is poor and supports only a few tough, weedlike plants: pale green clumps of sagebrush, snakeweed, and four-winged saltbush. There is almost no soil. Much of it is bare rock. Yet walk through the canyons that weave through the desert and you will find groves of cottonwood trees and pockets of green grass. Hanging gardens of ferns and small flowers lie hidden in grottoes and seeps in the canyon walls. Turn a corner in a nameless side canyon and you will find that the rocks are suddenly covered with drawings—pictographs and petroglyphs of giant, ghostlike figures in headdresses and kilts more than ten feet high, surrounded by images of antelopes and salamanders that scurry across the stones in shades of red, brown, and tan. Elsewhere the deserts are littered with the remains of abandoned villages and towns: the ruins of pueblos, kivas, and towers, the leavings of prehistoric people like the Fremont and Anasazi, who lived here more than seven hundred years ago. Here and there pieces of broken pottery and charred kernels of corn litter the ground—as if they had been abandoned only a few weeks before.

This world of slickrock canyons and abandoned cliff dwellings is all part of the Colorado Plateau, a vast, high desert that sprawls over four western states. Centered over a region known as the Four Corners, the place where the states of Arizona, New Mexico, Colorado, and Utah all come together, it is geologically, biologically, and culturally distinct from the high mountains and low deserts that surround it. It stretches all the way from the western edge of the Rocky Mountains in Colorado and New Mexico to the eastern edge of the Basin and Range country of Utah and Arizona. From north to south it reaches all the way from the Uinta Mountains in northern Utah near the Wyoming border to the Mogollon Rim in the volcanic highlands of central Arizona and New Mexico. Although it covers an area larger than New England, its total population is measured in thousands, not millions.

Its name is deceptive because the Colorado Plateau is not actually a single plateau but a broad belt of several dozen that spreads across the Four Corners region like a series of flat-topped tables. Their names are almost as colorful as the rocks themselves: Kaibab, Kaiparowits, Chaco, Wasatch, Tavaputs, Roan, Aquarius, and several others. In between them are broad desert valleys and high, solitary peaks, a landscape unlike any other in the western United States.

Here near the center of the Colorado Plateau the converging drainages of the Green and Colorado rivers have carved such an intricate network of canyons that the larger details of the landscape are almost impossible to see. Its surface seems almost entirely vertical: divided by successive walls of sheer smooth rock. It is only from the slopes of distant peaks that the larger plan and pattern of the landscape begin to emerge.

Fifty miles east of the canyon country here the Henry Mountains rise up out of the San Rafael Desert like islands, a range of five small cone-shaped peaks, the tallest reaching more than a mile up into the sky. From their top the view is panoramic, stretching for more than a hundred miles in every direction. High plateaus and peaks seem to line the horizon on every side. To the north is the long, low line of Book Cliffs. To the south a series of plateaus lead off like stepping stones toward the higher rise of the Kaibab Plateau, where the Colorado River twists and winds through the Grand Canyon. To the east you can see the snow-covered peaks of the San Juan Mountains that mark the edge of the Rockies. To the west, closer at hand, the high plateaus of the Wasatch Front seem to rise up like a solid wall, separating the plateau country below from the Basin and Range country of western Utah and Nevada—a corrugated landscape of high, narrow mountain ranges and long, flat, desert valleys. The surrounding rim of high country gives one the impression of standing at the center of a shallow bowl. Filling the foreground are broad, flat deserts, some covered with a green-gray haze of sagebrush and desert grasses, others with bare rock and shale. Here and there are other small ranges of islandlike peaks: the Abajos and La Sals to the east on the far side of the Colorado River, and the low moundlike rise of Navajo Mountain toward Lake Powell to the south. Between them the conical shape of Sleeping Ute Mountain is just barely visible on the edge of the horizon, sitting almost squarely on top of the imaginary point of the Four Corners.

Deep canyons cut through the rocks below, but they are scarcely visible, appearing only as thin seams in the rocks or patterns of dark shadows. What catches the eye instead is the broad, flat reach of the landscape. In spite of the sheer lines of the cliffs and the high rise of the plateaus, the rocks within them are almost perfectly horizontal—as flat and level as the day they were deposited. The ground is so bare and open that you can follow the path of a single layer of rock for miles.

The landscape here is enigmatic. The heart of the riddle, however, lies not within its deep canyons or its intricately carved arches and pillars of sandstone but in the flat, level layers of the rocks themselves. Although it lies between the mountainous reach of the Rockies and the rugged terrain of the Basin and Range Province, the land here is surprisingly calm and stable. Surrounded by belts of twisted and shattered rock, the rocks here have been shaped by nothing more dramatic than erosion.

The Colorado Plateau is an island in the sky, an isolated island of rock rising up out of the Desert Southwest. For nearly two billion years it was part of the earth's stable craton, or core, a place where sand and sediment accumulated, a solid block of rock untouched by the geological disturbances that shaped the land around it. Over the past five million years rivers and streams have cut down through its built-up layers of rock to leave the past out in plain view. In the face of a cliff or the edge of a high plateau you can see several hundred million years of time in the space of a few thousand feet of sandstone and shale. Flat, level layers of multicolored rock lie stacked up on top of one another like thick sheets of paper. You can read the earth's ancient history like a book. The Colorado Plateau is a landscape of levels.

Populations of plants and animals here are as far-reaching and varied as the area's built-up layers of rock. Surrounded by low deserts and high mountains, its flora and fauna are typical of neither mountain nor desert, but a quilted and confusing patchwork of both. Elevation here varies from less than 2,000 feet above sea level on the floor of the Grand Canyon to more than 12,000 feet above sea level on the top of high, solitary peaks like the La Sal Mountains of southeastern Utah. While the scrub and cacti of the canyon floors resemble those found in the deserts of northern Mexico and southern Arizona, the firs and aspens of the mountain slopes resemble those found in the boreal forests of Canada and Alaska. In between is a tangle of contradictions. The deserts here are cut by large rivers like the Colorado, Green, Rio Grande, and San Juan, which drain entire mountain ranges, but their water, for the most part, is out of reach and inaccessible, hidden away in deep canyons. While its isolated terrain has fostered the evolution of rare, new species of plants and animals, it contains others that have hardly changed at all in 2,000 million years of time—simple cells of blue-green algae or cyanobacteria that live on top of the thin desert soils.

The record of human history is no less deep here and offers its own tangle of contradictions as well. While the landscape seems almost uninhabited today, it contains the largest concentration of prehistoric ruins in the world. In the midst of an island of high, dry rock, prehistoric peoples like the Anasazi developed a sophisticated culture based on the cultivation of corn, beans, and squash. While Spanish friars and conquistadors had established a network of missions and villages here several years before the first Pilgrims reached Plymouth Rock, later waves of immigrants would leave this landscape almost untouched. Until the last century the desert canyons here were as unknown to outsiders as the mountains of the moon.

In the opening chapter of Willa Cather's *Death Comes for the Archbishop*, Jean Marie Latour, the newly appointed archbishop of New Mexico, is lost in the deserts of central New Mexico without map or compass, with only his sense of direction to guide him. "The difficulty was that the country in which he found himself was so featureless—or rather, that it was crowded with features, all exactly alike," Cather writes. "As far as he could see, on every side, the landscape was heaped up in monotonous red sand-hills, not much larger than haycocks." The desert is a land of infinite and often identical details, and it is that same overwhelming wealth of detail that makes the Four Corners region and the Colorado Plateau so hard to grasp. The region is crisscrossed by literally thousands of cliffs and canyons. In their faces you can see the rise and fall of ancient seas and the advance of ancient deserts. Travel to any corner of the region, north, south, east, or west, and you will find the same delicately tinted layers of sandstone, the same surreal shapes carved into the rocks, as the earth's ancient history is laid out in repetitive and often tedious detail. Climb to the top of a solitary mesa or butte and you will find yourself wandering through woodlands of pinyon and juniper as disorienting as the maze of canyons below. Out in the broad desert valleys that lie beyond the high rise of the plateaus and the narrow reach of the canyons you will suddenly find yourself lost in the midst of an endless sea of sagebrush and dry desert grasses that reaches as far as the eye can see. You can walk for days without seeing a single town—or at least any that have been inhabited anytime within the last five hundred years. Watch the ground closely, however, and you will begin to notice the signs of others: spearpoints and fire-cracked rocks from the camps and kill sites of Paleolithic hunters, a cairn of boulders and stones left by the Anasazi, a Spanish coin, or the wheel ruts of a Mormon wagon train. Things last a long time in the dry desert air, and the land here is littered with bits and pieces of the past, archaeological sites that record more than ten thousand years of human history—not a few thousand of them, but a few hundred thousand—so many stories and artifacts that it is almost impossible to bring them all into focus.

To understand a landscape like this, where a single glance can take in several hundred square miles of land and several hundred million years of history, one needs a definite plan or point of reference. The opening chapters of this book are points of departure, beginnings that lay out the broad sweep of the plateau country's unique geology, biology, and history—the rough borders of a verbal map. From here the text branches out in a series of journeys to the north, south, east, and west of the plateau country before returning back to the center. While each of those five sections is loosely focused on a particular region of the plateau, the divisions are as much temporal as spatial, taking one on a journey not only across the current surface of the plateau but through time as well, developing the history of both the land and its people side by side. Interwoven with all of this geologic and human history is the story of its varied plants and animals, from barren rock pools in the desert that fill with life at the first touch of rain to high alpine meadows that spend more than half the year buried in snow.

The place to begin is not here on top of a high peak but down on the floor of the Grand Canyon where the Colorado River cuts right down to the continent's core. A mile deep inside the earth you see the beginning of both life and time.

T W O

The Grand Staircase

THE CANYON WALLS rise right up from the river's edge in sheer black shapes, a narrow frame for the sky overhead. Bright points of light from the stars in the night sky dance across the surface of the Colorado River. A mile above, the canyon rim is buried beneath three feet of snow. Here on the floor of the Grand Canyon, however, the temperature is still well above freezing. The air is filled with the sound of rushing water as the river tumbles over the tangle of boulders and stones that marks the mouth of Bright Angel Creek. It is late January, and the canyon floor is all but deserted. In mid-summer several hundred pass by this point each day, a collection of hikers, rafters, and mule riders. In the early hours of a clear winter morning I am alone with the river.

The trail that leads down to the river's edge passes by the ruins of a small Anasazi pueblo. The square walls of its rooms and the circular shape of its kiva are visible in the starlight as I walk by. They seem to rise up right out of the rocks. The ruins are more than one thousand years old, leavings of the Anasazi, relatives of the same prehistoric pueblo peoples who built the soaring cliff palaces of Mesa Verde in southwestern Colorado and the sprawling cities and towns of Chaco Canyon in northwestern New Mexico. They did not last long here. This small cluster of rooms on the canyon floor may

have been inhabited for no more than fifty years. For all their antiquity, however, the Anasazi are only recent arrivals. The rocks on
which these simple ruins stand are nearly 2,000 million years old.
They date back to the earth's earliest history, a time when there was
almost no life on earth and the continents and ocean basins that now
divide its surface were only just beginning to take shape.

Sitting on the floor of the Grand Canyon looking up at the stars
is a good place to contemplate the earth's beginnings. Fifteen thousand million years ago, or so scientists believe, the universe was created by an explosion of gas known as the Big Bang. The force of the
explosion caused the universe to expand, inflating like a balloon, creating both space and time. At first that primitive universe contained
only radiation—light and radio waves alike. In time, however, these
tiny packets of energy were slowly transformed into matter, simple
elements like hydrogen that gathered together and ignited to form
stars. From time to time some of these glowing collections of material became unstable and collapsed inward to form supernovas, black
dead stars, concentrated points of matter where more complex elements of matter were created—carbon, oxygen, iron, and nitrogen.
Roughly 5,000 million years ago our sun was created from the collapse of one of these primitive stars. Surrounding it was a swirling
disk of other debris, clouds of matter left by the star's collapse. In
time eddies began to form within this swirling disk of debris, just as
eddies and whirlpools take shape in the swirling rapids of the
Colorado River as it winds through the Grand Canyon today. Drawn
together by their own gravitational pull, these swirling eddies of
debris began to collapse inward upon themselves, forming even
denser concentrations of matter—the beginnings of planets. Some
4,600 million years ago our own earth took shape from one of these
collapsing eddies of debris and began orbiting the sun.

In time the earth's surface began to take shape, differentiating
itself into continents and ocean basins—mobile pieces of relatively
solid crust floating on the more fluid layers of the earth's mantle,
changing and rearranging themselves with time. The rocks here on
the floor of the Grand Canyon date back to that initial formation of
the continent. They are some 2,000 million years old.

There are, of course, older rocks to be found on the earth's surface: the 3,500-million-year-old rocks of the Canadian Shield, for

example, that lies above the Great Lakes in central Canada. Others of equal antiquity can be found in the deserts of Australia and Africa. What is so unique about these ancient rocks in the Grand Canyon is not so much age as context. While those on the canyon floor may be as much as 2,000 million years old, those at the canyon rim are no more than 250 million years old. In between is a nearly continuous record of the earth's ancient history. In the space of some five thousand feet of rock, more than a billion years of the earth's ancient history is laid open to plain view.

The next day the sun was bright and warm on the canyon floor. Walking alongside the river you could see Joshua trees and tough, straggly clumps of mesquite clinging to the cliffs above. Plants from the low, hot deserts of southern Arizona, they have followed the deep cut of the river northward into the high, cold deserts of the Colorado Plateau. In places you could see all the way to the canyon rim and its snow-covered forests of ponderosa pine. Below it the rocks that lined the upper reach of the canyon seemed to be laid out in flat, brightly colored layers of red, white, brown, and tan.

Down here in the inner gorge, however, the rocks are not flat and colorful but twisted and black. In places they seem to have been folded almost double. Here and there veins of white rock shoot through the canyon walls. They look like bolts of lightning frozen in stone. The rocks here are so hard that the river has been unable to cut anything more than a narrow slot. In places the walls of the inner gorge are less than a half mile apart. Above it the brightly colored layers of rock that rise up toward the canyon rim are soft and easily eroded. The steady retreat of cliffs and the gradual wearing away of rock have widened the upper reaches of the Grand Canyon to the point where its North and South rims are now more than twelve miles apart.

The black rocks that line the river here are part of a formation known as the Vishnu Schist. Metamorphic rocks, they have been exposed to intense heat and pressure. Look closely and you can see how its intricate layers seem to be made up of thin, needlelike layers of rock—flakes of black and white mica, feldspar, and quartz. Geologists believe that these rocks were once the roots of an ancient mountain range. Similar twisted layers of rock may lie several miles beneath the Himalayas or Alps today. They have been so altered and deformed, however, that their exact age and original features are hard to determine. Their chemistry suggests that they were once marine, perhaps a collection of ancient sandstones and shales deposited on the floor of one of the earth's original seas. Here and there patterns of cross-bedding and faint ripples are still visible— marks of moving water. The only signs of life are traces of calcium silicate that suggest the presence of lime-secreting algae, primitive single-celled organisms that were quite possibly the first forms of life on earth. They may have been deposited as long as 2,000 million years ago, but that date is largely speculative.

But while the age of these ancient rocks is uncertain, the timing of the metamorphism that altered them is more clear-cut. Dating reveals that the metamorphic event that reshaped these ancient rocks took place roughly 1,700 million years ago. That is an interesting date to geologists because beginning about 1,600 million years ago the craton, or core, of North America grew by more than 50 percent as pieces of the earth's crust were welded and sutured together by collisions and episodes of mountain building. The Vishnu Schist that lie at the floor of the Grand Canyon may have taken on their present shape from an ancient episode of mountain building linked to the very beginnings of the North American continent. After this initial buckling and folding, at least two other later metamorphic events would lace these dark rocks with dikes and sill—veins of white-hot molten rock that shot through the older rocks above.

Roughly 1,300 million years ago these twisted rocks from the core of an ancient mountain range were uplifted and planed almost flat by erosion as the mountains that once rose above them were completely worn away. By 1,000 million years ago they had begun to move inland across this newly flattened plain, depositing a thick layer of sandstone and shale known as the Grand Canyon Series. Originally deposited in flat, level layers, another episode of uplift would tilt them to an angle of ten to twenty degrees. It would be the last bit of geologic activity this region would see for several hundred million years. Afterward the sea would move in again and plane these newly deposited layers of marine rocks flat as well. Although you can still see the tilted layers of rock from the Grand Canyon Series quite clearly from places like Desert View farther up river, here near the center of the canyon they were almost completely eroded away. Left behind was a broad flat plain, setting the stage for more than 500 million years of quiet deposition that would cover the continent here with more than two miles of rock—the solid block from which the Colorado Plateau and the Grand Canyon would later be cut.

I walked out of the inner canyon that afternoon, up through the bent black layers of rock that line the river to the brightly colored layers of the larger canyon above. In midsummer the late afternoon heat of the Inner Gorge is deadly, with temperatures that climb above 110 degrees. In midwinter, however, the warmth of the sun was a welcome contrast to the cool shade of the canyon floor.

The contrast between the ancient rocks from the continent's core that line the inner canyon and the younger marine rocks above is plain to see. The hard rocks below form a stable base for the softer layers above. Higher up, retreating cliffs have left behind a broad, benchlike platform known as the Tonto Plateau. It runs the length of the canyon along both the north and south sides of the river, a sharp dividing line between the twisted rocks from the continent's core below and the flat layers of rock from ancient seas and shorelines above. Climbing out onto the Tonto from the confined and narrow reach of the Inner Gorge is like stepping into another world. Suspended between the rim and canyon floor, the ground seems so broad and open that it feels like one could fly right off into space. Looking back down into the canyon, you can see the record of nearly a billion years of time.

In the legends of the Southwest's Pueblo people, the earth's ancient history is divided into a succession of worlds—former earths that lie beneath the surface of our present one through which life and man have progressively emerged. According to nearby tribes like the Hopi, whose villages cluster atop a series of mesas some eighty miles southeast of the Grand Canyon, in the beginning of time there was only *Tokpella*, endless space, a world without light, sound, or substance. In time, *Tawa*, the Sun Spirit or Creator, gave shape to the First World, a primitive earth that was inhabited only by a race of insects who dwelt in a deep cave. Outside the confines of this early world, however, were others—each one progressively richer and more like the present. We are, they say, currently living in the Fourth World. Others may lie outside it.

Scientists view the earth's history as a similar succession of worlds. Geologists divide the vast reach of the earth's history into successive eras, the Precambrian, Paleozoic, Mesozoic, and Cenozoic, the last three representing the intervals of ancient, middle, and recent life respectively. The breaks between these successive eras of time are punctuated by dramatic extinctions that set the stage for successively more complex forms of life.

The Precambrian begins with the earth's creation some 4,600 million years ago. The first signs of life, however, did not appear until some 2,500 million years ago—simple cells of blue-green algae that grew on the seafloor in large mats known as stromatolites. For more than 1,000 million years they may have been the only form of life on earth.

The start of the Paleozoic was marked by the sudden appearance of marine invertebrates some 570 million years ago. Primitive relatives of modern snails, clams, corals, and starfish, they left large, readily identifiable fossils behind worldwide. The Paleozoic saw the appearance of the earth's first fishes, amphibians, and reptiles and the appearance of the first multicellular land plants as well: forests of giant ferns and primitive trees. Some 245 million years ago, however, it all came to a close with a massive extinction that wiped out 90 percent of the earth's species of marine invertebrates, setting the stage for the Mesozoic and the rise of dinosaurs.

Here in the Grand Canyon the major breaks of the landscape coincide neatly with those of ancient geologic time. While the top of the Tonto Plateau coincides with the start of the Paleozoic time and the sudden appearance of marine invertebrates, those of the canyon rim mark its close and the massive extinction that set the stage for the Mesozoic and the rise of dinosaurs.

Standing on the Tonto at the base of Paleozoic time you can see up and down the canyon for miles. Terraced layers of red, white, and tan rock lead up to the canyon rim, as flat and level as the day they were deposited. Hot desert plants like the mesquite and Joshua trees of the canyon floor give way to cold desert plants like sagebrush and Mormon tea. Here and there the tall, dry stalks of flowers from solitary agaves rise up from the ground like feathers. Stands of cottonwood trees and green thickets of cattails mark the site of springs and seeps at the base of the cliffs. In places small streams and drainages have cut winding canyons into the broad bench of the Tonto—pathways to the river below. Along their edges you can see blocks of rock the size of large houses and small city blocks that have broken off the sides of cliffs and canyon rims and tumbled several hundred feet below. Out on the edge of the Tonto you can hear the roar of rapids on the Colorado River more than a quarter mile below. Up above, the retreating walls of the outer canyon have cut back into the rocks in great, curving amphitheaters, leaving points of rock projecting out into the empty space of the canyon like the prows of giant ships.

I spent the rest of the day walking along the Tonto Plateau, watching the clouds drift across the sky to send patterns of sun and shade dancing across the rocks. By late afternoon a winter thunderstorm had settled into the canyon, bringing swirling clouds of snow that melted as soon as they touched the ground. The wind whistling along the canyon walls sounded like the surf of ancient seas.

For nearly 300 million years, the land here was part of a broad costal plain alternately above and below sea level. As ancient seas rose and fell over what is now the desert Southwest, the landscape here was covered with alternating layers of marine and terrestrial rocks. The base of the Tonto is made up of a formation of rocks known as the Tapeats Sandstone. A brown-gray cliff-forming sandstone, its lowermost layers are laced with small pebbles and stones that gradually give way to fine sand. Bedding patterns within the sandy layers suggest that it was deposited somewhere near shore. There are also abundant signs of life: fossils of trilobites and brachiopods, hard-shelled marine invertebrates, part of the explosion of life that marks the base of the Paleozoic. Layers of Tapeats Sandstone are older and thicker to the west, suggesting that the ocean invaded from that direction and became gradually deeper with time. That trend is continued by the rocks that lie above it—the Bright Angel Shale and Mauv Limestone.

The Bright Angel Shale varies from greenish gray to buff-colored. Its greenish tinge comes from glauconite, a mineral precipitated from seawater and found only on the outer reaches of the continental shelf—the boundary between the shallow reaches of the coastal ocean and the depths of the open sea. Soft and easily eroded, it forms the wide bench of the Tonto. Up above it the Mauv Limestone appears, a record of still deeper water. Harder than the soft shale layers below, it begins the first of a series of steep climbs up the face of the canyon walls. In time, however, that rising sea would retreat, exposing these deepwater rocks to erosion. By the time the tower walls of Redwall Limestone that dominate the outer walls of the canyon appeared, more than 150 million years of time had disappeared.

From the river's edge to the canyon rim, that pattern of rising and falling sea level is repeated no less than five times in the walls of the Grand Canyon. Changing environments left behind rocks of different colors: white sandstones from beaches and dunes, red shales and muds from swamps and coastal lagoons, as well as cream- and tan-colored layers of limestone from deeper water. The list of formations and rock layers reads like the index of a textbook in introductory geology: the Redwall Limestone, Supai Formation, the Hermit Shale, Coconino Sandstone, Toroweap Formation, and the Kaibab Limestone.

These repetitive transgressions of sea level have led geologists to surmise that the Grand Canyon region during Paleozoic time was part of a broad and almost featureless coastal plain that extended for hundreds or even thousands of miles in almost every direction— seldom more than a few feet above or below sea level. To the west was open ocean. California and Nevada were nowhere to be found. To the east was a broad flat plain. The Rockies would not rise for several hundred million years. To the north and south this flat world was seemingly limitless as well. Units like the Redwall Limestone, the five-hundred-foot-plus thick rampart of rock that forms the sheer, prominent cliffs of pale, reddish-tan rock in the outer walls of the canyon, are found as far north as Canada and as far south as Mexico.

That broad reach of the rocks is spectacular, but not entirely unexpected. The rocks exposed here in the walls of the Grand Canyon do not disappear once they dive beneath the rim, but extend over thousands of square miles of the Intermountain West—the broad, high, dry reach of mountains and plateau-studded land that lies between the Rockies and the Sierras. Peel back the Rocky Mountains today and you will find the same rocks that line the walls of the Grand Canyon a mile or more beneath the surface. Some run eastward into the Great Plains. They blanket the base of the Colorado Plateau as well. You can find the same sheer cliffs of Redwall Limestone that rise sbove the Grand Canyon's Tonto Plateau, for example, exposed near the Utah-Wyoming border in the Canyon of Lodore cut by the Green River, and the same Kaibab Limestone that makes up the canyon rim in central Utah's San Rafael Swell. The rocks here are all part of the foundation of the Colorado Plateau.

During the time these rocks were being deposited, the ancestral Colorado Plateau was located roughly on the equator or perhaps even slightly south of it. It was also rotated some ninety degrees. What now faces west once faced north. While earthquakes along the San Andreas Fault and fiery volcanoes in the South Pacific give one the impression the earth's surface is in constant motion, for more than 300 million years the land here hardly moved at all. Sand and sediment slowly piled up layer upon layer, the way dust accumulates on the flat surfaces of tables and shelves in a tightly closed room. Compressed by their own weight, these deep piles of sand and silt were slowly transformed into solid rock. Other continents, however, were already in motion, drifting across the earth's face to finally gather together along the equator, centered around the edges of ancient Africa. By the close of the Paleozoic some 245 million years ago the earth's continents would be joined together in a single landmass, an ancient supercontinent known as Pangea. While the continents were joined together, the earth's seas and oceans were hit by a massive extinction that wiped out some 90 percent of the existing species of marine invertebrates—the same hard-shelled animals whose appearance had marked the opening of the Paleozoic some 320 million years before.

Geologically speaking, however, the region was still calm and isolated, a place where rocks and sands quietly accumulated. In time this ancient gathering of continents would begin to break up. Afterward North America would begin moving as well, heading northward and carrying the ancestral Colorado Plateau region with it through a succession of tropical rain forests and sandy deserts to what we now think of as its final position roughly halfway between the equator and the pole.

• • •

I spent the night camped out on the Tonto Plateau with nothing but the stars overhead for a roof. I awoke the next morning to the sound of falling rock. Looking up I could see blocks of pale, tan-colored sandstone from somewhere up near the rim tumbling down the canyon wall. They seemed to gather speed and debris as they fell, leaving a plume of dust in their wake, like the cloud of snow that follows the path of an avalanche down the flanks of a high peak. The canyon had grown that morning.

Early geologists, relying more on speculation than evidence, assumed that the canyon was the product of some cataclysmic event—a violent earthquake, for example, or a massive flood. In point of fact, however, the canyon was created by nothing more dramatic than the steady pull of gravity and erosive power of running water—millions of small landslides like the one that had woken me from sleep that morning and the steady cut of the river as it runs across the rocks below. First-time visitors to the Grand Canyon often assume that the canyon itself is very old. But while the rocks exposed on the floor of the canyon may be as much as 2,000 million years old, the canyon itself may be no more than 20 million years old and possibly as little as 5 million years old. Sometime over the past few million years movements beneath the earth's surface began to lift the Colorado Plateau region upward by more than a mile. As the land rose up, its rivers cut downward, exposing the buried rocks below.

After a pot of tea and a quick breakfast I began the long climb up to the canyon rim some three thousand feet and 300 million years above. Near the base of the Tonto Plateau I could see the fossils of trilobites and primitive clams. As I headed up the cliffs, the layers above revealed not only the increasing depth of the earth's surface here with time but its increasing complexity as well: the fossils of primitive corals and fish, followed by those of terrestrial plants and animals near the canyon rim—not mere bones or shells but the footprints of ancient insects and amphibians and the small craterlike depressions left by raindrops that fell more than 250 million years ago.

The walk up to the canyon rim from the Tonto is staggered, a series of steep climbs punctuated by narrow benches. Differences in hardness between the successive layers of rock have given a stairstep appearance to the canyon walls. Softer layers erode into shallow slopes and benches, undermining the harder layers above them that crumble and fall away in blocks leaving sheer cliffs behind. Plants and trees change as well: The desert of sagebrush and dry grass that covers the Tonto gives way to thickets of pinyon and juniper followed by forests of ponderosa pine near the canyon rim. Above the steep cliffs of the Redwall Limestone the trail is covered with ice and an increasingly thick blanket of snow.

Up on the canyon rim the rocks suddenly come to an end with the Kaibab Limestone, the 250-million-year-old cap of the Grand Canyon that marks the close of the Paleozoic time here. The same erosion that has cut the Colorado River's deep canyon here has also stripped away younger rocks that once lay above what is now its rim. If those younger rocks had not been worn away, what is now the canyon rim would be roughly its midpoint as another mile-thick sequence of rocks would extend upward above your head into the sky. Those rocks, however, have not vanished completely. Standing on the South Rim of the Grand Canyon you can still see them by facing north and looking across the vast, deep reach of the canyon toward the horizon. They appear in the distance as successive banks of multicolored cliffs rising behind the high walls of the North Rim: the Chocolate, Vermillion, White, Grey, and Pink Cliffs. They seem to rise up out of the ground almost like stairsteps running northward into Utah. The clearly exposed layers of rock here are all part of what geologists call the Grand Staircase—a nearly complete record of the earth's ancient history that begins with rocks more than 1,000 million years old on the floor of the Grand Canyon and continues farther north through a progression of successively higher plateaus and cliffs to thin layers of soil and glacial debris only a few thousand years old. A mile above the desert in the high plateaus of southwestern Utah the surface of the Colorado Plateau comes face-to-face with the present.

The snow has been falling for two days now. It drops down out of the overcast sky in thick, wet flakes, falling into Zion Canyon to cover the desert with snow. Gray clouds drift and swirl through the low sky, hiding the canyon walls from view. In the dim storm light the shallow waters of the Virgin River seem black and swift. Cottonwoods line the river's bank. There is almost no wind, and the snow lies on their outstretched branches in thin, narrow layers.

A thousand years before, the canyons here were home to prehistoric pueblo people like the Fremont and Anasazi. Later they would abandon the deserts here just as they had in the Four Corners region farther to the east. A century or more after they had vanished, a tribe of nomadic desert peoples known as the Paiute began moving through the area, hunting and gathering, collecting roots and seeds. They left Zion Canyon itself untouched, however, believing it to be the realm of dangerous gods and spirits: *Wai-no-pits*, the evil one, who invaded their camps with sickness and death, and *Kai-ne-sava*, the God of Fire who set bolts of lightning dancing across the canyon's high pillars of rock. When the first Mormon scouts reached the area in the mid-1800s, their Paiute guides would venture no further than Oak Creek, near the canyon's mouth.

Later, Mormons would settle in the isolated canyons and plateaus of southern Utah and eventually Zion Canyon itself, farming and grazing their herds on the canyon floor of the area until its establishment as a national park in the early 1900s. They saw it as a place of grace and power. "These mountains are natural temples of God," said Isaac Behunin, one of the canyon's first settlers. "We can worship here as well as in the man-made temples in Zion, the biblical heavenly city of God. Let us call it Little Zion."

The Virgin River slices through the plateau country here just as the Colorado River has sliced through the Kaibab Plateau farther south in Arizona to create the Grand Canyon. Its look, however, is entirely different. The canyon floor here is shaped like a funnel. At its mouth near the edge of the park it is more than a mile and a half wide. Less than ten miles upstream it shrinks to a narrow sixteen-mile-long slot between curving walls of sandstone whose two-thousand-foot-high walls are often no more than an arm's breadth apart—a stretch of the river known as the Zion Narrows. Out in the wider reach of the canyon below the details of those sandstone walls are out in plain view. Sheer cliffs of red-tinted sandstone rise up more than two thousand feet above the canyon floor, a red rock Yosemite, their tops carved and worn by water and ice into fantastic rounded shapes—soaring pillars and cones of rock with names inspired by the visions of Mormons and Paiutes—the Temple of Siwanava, the Great White Throne, the Court of the Patriarchs, Angels Landing.

With the snowfall the temperature within the canyon hovers within a few degrees of freezing. Streams from the deeper fields of snow and ice above the rim cascade down the canyon walls in thin, narrow ribbons. In places they have begun to freeze, creating thin columns and needles of ice ten, twenty, and thirty feet long. From time to time you can hear them crack and break—followed by the roar of falling blocks of ice as they tumble to the canyon floor. Here and there trails follow narrow fissures and side canyons up into the rock, passing through sheltered groves of Douglas fir and hanging gardens of moss and fern up to an open rolling reach of sandstone and slickrock quilted with dry, low forests of pinyon and juniper.

The rocks exposed here in Zion Canyon and the national park that surrounds it take over the earth's history roughly where the Grand Canyon left off. The Kaibab Limestone, the same late Paleozoic limestone that holds up the rim of the Grand Canyon, lies just below the surface here. Above it the rocks are all Mesozoic, from the age or interval of middle life. Just as life in the oceans underwent an explosion of diversity in the Paleozoic, life on land would undergo an explosion of diversity during the Mesozoic, filling the earth with startlingly new and diverse forms of life. Most spectacular and perhaps best known were the dinosaurs—their name a combination of Greek words meaning "monstrous lizards"—giant reptiles, some more than a hundred feet long and thirty feet high, that roamed the earth between 65 million and 220 million years ago. They left their bones and tracks scattered across the surface of the Colorado Plateau—the skeletons and even footprints of giant animals like the stegosaurus and brontosaurus.

While the ancient rocks of the Grand Canyon record the initial assembly and buildup of the continent, those here in Zion Canyon record its gradual drift northward from the equator as the supercontinent of Pangea, that union of the earth's continents into a single landmass that marked the close of the Paleozoic some 245 million years ago, began to break up. That movement is not recorded in dynamic features like faults or folds or mountain ranges but in the composition of the rocks themselves: changing fossils and patterns of sand and mud that record a steady progression from tropical rain forests and dry sandy deserts to its present location in the Northern Hemisphere.

The story of the Mesozoic begins with the chocolate-colored rocks of the Moenkopi Formation found in cliffs alongside the Virgin River just outside the southern boundaries of the park. A collection of red beds, shales, siltstones, sandstones, and limestones, the Moenkopi varies in color from pink to chocolate brown. Its rocks are roughly 240 million years old. The land here was still part of a broad coastal plain, underwater at first but becoming generally drier with time. The blood red color of its uppermost layers is a result of hematite—a form of iron oxide—suggesting that the rocks and sands here were exposed to open air when they were deposited.

Streams would eventually wash away the upper layers of the Moenkopi, leaving behind a blanket of gravel known as the Shinarump Conglomerate. Like the broad coastal plain that preceded it, this erosional surface was vast. Today pebbly layers of the Shinarump cover some one hundred thousand square miles of the Intermountain West.

The pebbly conglomerates, in turn, lie at the base of the Chinle Formation, the remains of a vast tropical river that stretched all the way from the Texas Panhandle to Nevada some 215 million years ago. Neither the Rockies nor the Sierras existed at the time, and the river flowed westward right through the center of what is now the Colorado Plateau, possibly all the way to the Pacific. One of the most varied formations in the Southwest, the Chinle is made up of shales, sandstones, and limestones that vary from gray to purple to white or even green. Soft and easily eroded, the rocks of this ancient river system make up the badlands terrain of the Painted Desert in northeastern Arizona. Both the types of rocks deposited and the fossils they contain paint a picture of a broad, shallow river floating across a gently sloping plain where meandering streams fed a network of marshes and ponds. In places tangles of driftwood were left behind on sandbars, preserved today as pieces of petrified wood at places like Petrified Forest National Monument. Although the land was still within a few feet of sea level, fossil plants and animals found in the Chinle suggest that fresh water was abundant, leading geologists to believe that the area had drifted northward into a belt of heavy rainfall north of the equator—perhaps somewhere near the latitude of present-day Sierra Leone in Africa.

That pattern of meandering streams continues upward through the next two layers of overlying rock, the Moenave and Kayenta formations—although both the climate and as a result the character of these meandering streams had changed dramatically. By the time the red and mauve rocks of the Kayenta Formation were deposited some 195 million years ago, rainfall seems to have become more seasonal. Stream- and lake-bed deposits laced with fossil mud cracks suggest that the area was seasonally dry. At the time these rocks were deposited the area seems to have been located at roughly the same latitude as the African Sahel, a drought-plagued landscape of rainy summers and dry winters on the edge of the Sahara, centered over the West African nations of Senegal, Mali, Niger, and Chad. Over the next few million years, however, the continent's steady northward drift would carry it beyond this belt of seasonal rain, transforming its surface into a barren desert.

By 190 million years ago the Colorado Plateau lay at the center of a vast sand sea—a dry, waterless desert covered with rolling dunes of sand. Its steady drift had carried it to roughly the same latitude as that occupied by the Sahara today. The sand from those ancient dunes makes up the two-thousand-foot-high walls of Navajo Sandstone that tower over the floor of Zion Canyon. Though the rock is white on top, staining by groundwater has given the layers exposed in the canyon walls a reddish tint. Its surfaces are so smooth and sheer that they are often mistaken for granite. Look closely, however, and you can see the steady action of ancient winds in the giant patterns of cross-bedding that climb up the canyon walls, creating a crosshatched pattern in the rocks like the threads in a herringbone tweed. The finely detailed layers were created as individual grains of sand blew over the wavelike shape of ancient dunes, leaving flat, tilted layers of sand leaning against one another. Analysis of their orientation suggests that the prevailing winds blew out of the northeast, just as they do in the tropical deserts found in the latitudes of the Sahara today.

Deposits of Navajo Sandstone stretch all the way from central Wyoming to eastern Colorado and southward into New Mexico and

Arizona as well. It covers an area of some 150,000 square miles, an area roughly the same size as the present-day Sahara. On the Colorado Plateau it can be found everywhere from the walls of Zion Canyon to the shores of Lake Powell. Built up out of loosely cemented layers of windblown sand, on the tops of plateaus it weathers into rounded, curving white mounds so smooth and surreal they look almost as if they had taken shape on a potter's wheel.

For all its dryness, however, the land here was still only a few feet above sea level while these sandstones were being deposited. Continued drift would eventually carry the land back into rain. By 170 million years ago seas and streams would begin to move back and forth across the surface of these ancient dunes, leaving behind a succession of tan and gray limestones and red clays—the Temple Cap and Carmel formations that lie atop the sheared walls and monoliths of Navajo Sandstone found here in Zion Canyon. In places these shallow marine rocks are laced with oolites—tiny balls of algae that have been transformed into limestone. Formed by the steady wash and roll of waves over shallow water, similar features are found in the tropical lagoons and banks of the Caribbean today.

Erosion would then wear these rocks down again, beveling the nearshore and shallow-water rocks of the Carmel. No new rocks would be deposited here until the arrival of the Dakota Formation some 100 million years ago near the close of Mesozoic time. Made up of tan-colored sandstones and laced with the fossils of terrestrial plants and a variety of clams known as pelecypods, layers of the Dakota can be found as far away as the Great Plains. Although it was still barely above sea level, the land here was located at roughly its present latitude by late Mesozoic time. Like the layers of the Chinle and Kayenta formations below, the younger rocks of the Dakota are laced with stream deposits, but the direction of flow is different. Water was no longer flowing westward across the surface of the ancient Colorado Plateau but eastward. To the west California and Nevada were beginning to take shape, and collisions and disturbances associated with their arrival had created a series of highlands to the west. The modern world was beginning to take shape.

If the ancient rocks exposed in the walls of the Grand Canyon are part of the plateau country's hidden foundation or core, those exposed here in Zion Canyon are its visible walls and sides. Just as they cover the older rocks of the Grand Canyon below, they too were in turn once covered by younger rocks as well. Over much of the Colorado Plateau today, however, those younger rocks have been all but washed away—preserved only in the region's high plateaus and deep basins—leaving these rocks from the middle reach of the earth's history exposed at the surface.

You can find the same rocks exposed here in the walls of Zion Canyon coloring the face of canyons and mesas all across the Colorado Plateau. They make up much of the scenery not only at such well-known places as the Glen Canyon National Recreation Area and Canyonlands National Park in southeastern Utah, but also lesser known places like the San Rafael Swell in central Utah and the vast reach of the Navajo Reservation in northern Arizona. Bits and pieces of it extend far to the east as well, underneath the Rockies and out into the Great Plains. At the time these rocks were being deposited, however, the Colorado Plateau was still indistinguishable from much of the land that surrounded it. After more than 400 million years of slow and steady deposition it was still part of a broad coastal plain, a place where shallow seas and deserts slowly advanced and retreated. It reached almost uninterrupted from Canada to Mexico.

In spite of the landscape's sameness and continuity during this middle time, the Mesozoic rocks seen here in the walls of Zion Canyon vary in thickness from place to place due to differing rates of erosion and deposition. In some spots they thin out or even disappear completely. Elsewhere they are joined by other layers of similarly colored and textured rocks: the salmon and tan-colored layers of the Wingate, Entrada, and Curtis sandstones of southeastern Utah and the improbably colored shales and clays of the Morrison Formation that spread across northern Utah and Colorado. To the casual eye these varying units often seem almost indistinguishable, appearing like variations on a common theme, as if geology here were cyclical, repeating the same patterns over and over again for tens of millions of years. Even with these regional elaborations, however, the basic layers of the Mesozoic seen in Zion Canyon are still there to help tie the landscape together—multicolored layers of Kayenta, Navajo, and Dakota sandstone, the center and heart of the vast block of rock from which the Colorado Plateau would later be cut.

The Mesozoic drew to a close like the Paleozoic before it with a massive extinction that wiped out the dinosaurs and cleared the earth's surface for the rise of mammals and other new forms of life in the Cenozoic. In the space of some 180 million years the steady northward drift of North America had carried the Southwest through some twenty-seven degrees of latitude, a journey of nearly two thousand miles. In spite of their long traverse, the rocks here were still flat and relatively undisturbed. Soon after the start of the Cenozoic, however, the once-level surface of the Colorado Plateau would be creased with faults and folds. The modern world was still waiting to arrive.

It is a few minutes before six on an early morning in January. The blue hour. The approaching sunrise is little more than an hour away, marked by a thin line of indigo on the edge of the horizon. Overhead the sky is pure black and dusted with stars. The moon has been down for several hours and the temperature has fallen to fifteen below zero. The air is sharp and clean. As I snowshoe along the edge of the plateau in the early morning light it stings my eyes and nose. In the clear, cold air the desert is empty of both life and sound. Down below the floor of Bryce Canyon is covered with fins and spires of rock. They march out from the base of the cliffs in endless rows—blue and black shapes in the starlight, the final steps of the Grand Staircase.

The park's name is somewhat deceptive, for Bryce Canyon is not really a canyon at all but more of an amphitheater of brightly colored rock carved into the edge of the Paunsaugunt Plateau. Part of the high plateaus of the Wasatch Front, it lies near the southwestern corner of the Colorado Plateau. At an elevation of nearly nine thousand feet the ground here is not covered by a bare desert of sagebrush or a scrubland of pinyon and juniper but by a mountain forest of spruce and fir. Here and there the twisted trunks of thousand-year-old bristlecone pines rise up out of the snow to overlook the desert below.

In summer the temperatures here can climb to a hundred degrees, but at nearly two miles above sea level, winter storms leave far more than a dusting of snow. It lies in deep drifts between the thin, straight trunks of the trees that cover the top of the plateau.

Along the canyon rim the wind has created twenty- and thirty-foot-deep cornices of snow that hang out over the edge of the rocks like frozen waves.

The Paunsaugunt Plateau is a dividing line of sorts. Rain and snow falling on its eastern flank flows to the Paria River and eventually on to the Colorado to speed through the Grand Canyon as it heads toward the Gulf of California and the sea. Moisture falling on its western flank, however, never reaches the Pacific. Instead it flows toward the Sevier River and heads out into the land-locked deserts of western Utah, evaporating in the narrow desert valleys and salt flats of the Great Basin that borders the western edge of the Colorado Plateau.

By seven o'clock the sky is a solid wall of orange. Eighty miles away the rounded shape of Navajo Mountain is backlit by the glow of the early-morning sky. Beyond it the successive lines of plateaus and cliffs lead off all the way to the edge of the horizon.

As the sun rises into the sky, the landscape begins to glow. Banks of snow on the plateau's edge shift from blue to red, then orange, yellow, and white as the day begins to take shape. Down below the blue-gray landscape of early morning explodes into color. The vague shapes of spires and fins become bright pillars of improbably colored rocks—layers of orange, pink, and white rock, their tops capped by deep piles of snow. Between them are narrow cuts and cracks in the fabric of the rock, a maze of pathways that lead downhill to the edge of the snow- and sage-covered valley below.

The rock layers exposed here in the high cliffs of Bryce Canyon are all Cenozoic, from the era of recent life, the brightly colored remains of ancient lakes. They are less than 60 million years old. Those in the valley below, however, are still Mesozoic, a continuation of those found farther south in Zion National Park. The same Dakota Sandstone that marks the uppermost reach of the rocks there lies just beneath the surface. On top are soft and easily eroded layers of Mancos Shale—known here as Tropic Shale—and a collection of sandstones known as the Mesaverde Group. As their name implies, the same rocks make up the sloping tableland of Mesa Verde farther east, where the Anasazi built some of their most striking and beautiful cliff dwellings. To distinguish between the plateau and the rocks that share its name, geologists run the two words together when speaking of the widespread layers of rock that make up the mesa itself.

The Mancos Shale represents the remains of a vast interior seaway—the Mancos Sea—that stretched across western North America all the way from the Arctic Ocean to the Gulf of Mexico some 80 to 90 million years ago. Soft and easily eroded, here on the Colorado Plateau it typically forms a badlands terrain of low cliffs and hills like the broad open deserts of the Four Corners area that stretch southward from Mesa Verde in southwestern Colorado toward Shiprock and Gallup or the barren clay hills of south-central Utah that surround the Henry Mountains. Above it sandstones of the Mesaverde Group record the gradual retreat of this inland sea.

The history of the brightly colored Cenozoic rocks above are different still, not part of an ancient inland sea or a chain of swamps and marshes, but the remains of a freshwater lake that covered the ground here some 55 million years ago. Known as Lake Flagstaff, it is some 250 miles long and 75 miles wide, spreading north and south over what is now a belt of high plateaus on the western edge of the Colorado Plateau. Its size and shape were roughly identical to those of Lake Erie today. From the Bryce Canyon area it stretched northward almost all the way to Salt Lake City. You can find the brightly colored rocks of this ancient lake, known originally as the Wasatch Formation, but now subdivided into as many as four separate units, spread up and down the spine of the high plateaus that run through central Utah like the icing on a cake. Farther north they were overlain by the limestones and sandstones of Lake Uinta, yet another freshwater lake from Cenozoic time that reached northward into Wyoming. Here to the south these layers of lake rock were covered by flows of volcanic rock that erupted out of the ground between 10 and 25 million years ago—a protective cap for the softer rocks below, shielding them from erosion. Looking across the rolling surface of the Paunsaugunt Plateau you can see how these black basalts cover the brightly colored rocks below in the higher rise of Bryan Head to the west. Fifteen thousand years ago its surface was scored by glaciers.

These brightly colored rocks on the flanks of Bryce Canyon mark not only the final step of the Grand Staircase but a turning point in the history of the Colorado Plateau as well. When the gray and brown rocks of the Mancos Shale and Mesaverde Group below were being deposited, the surface of the Colorado Plateau was coherent and whole—barely distinguishable from much of what is now the Intermountain West—a shallow and stable place where the sands and sediments were deposited just as it had been some 500 million years before when the rocks that line the Grand Canyon's Tonto Plateau were being deposited—the fossil-laden layers of the Tapeats Sandstone, Bright Angel Shale, and Mauv Limestone. Look around the edges of the Colorado Plateau and you will find that the soft rocks of the Mancos Shale and Mesaverde Group all but encircle it—found in the flanks of the Book Cliffs, Grand Mesa, Mesa Verde, Black Mesa and the Kaiparowits and Wasatch plateaus. Not only do they rim the edges of the plateau, they once extended across it in a solid sheet as well. You can find traces of these blanketing sandstones and shales scattered all over the Colorado Plateau like the highwater mark left by a river running at flood.

After these soft brown and gray rocks were deposited the geologic history of the Colorado Plateau region would no longer be dominated by the deposition and accumulation of rocks, but by their disruption and removal. By the time the brightly colored layers of the Wasatch Formation were being deposited, an episode of mountain-building known as the Laramide orogeny was already beginning to reshape the Intermountain West, giving rise to the Rocky Mountains farther east and rippling the surface of the Colorado Plateau with a series of faults and folds—dividing its once-uniform surface into a series of isolated basins and highlands. Surges of volcanism that swept through western North America after the close of the Laramide orogeny largely bypassed the landscape here as well: defining its edges with lava fields and cinder cones and volcanic seams like the Rio Grande Rift that defined its limits even further. While eruptions were frequent around the edges of the plateau, within its center they left nothing more than small, isolated flows of lava and solitary peaks like the Henrys, La Sals, and Abajos, which rose up out of the ground like bubbles without even breaking the surface.

Ten to 20 million years ago the land to the west and south would begin to stretch and thin as the Basin and Range Province of Nevada, western Utah, and the southern reaches of Arizona and New Mexico began to take shape. In the midst of all this activity the Colorado Plateau region was once again left all but untounched. Five million years ago, when the San Andreas Fault began to slice through coastal California, the Colorado Plateau began rising up toward the sky, uplifted by more than a mile. While surrounding deserts and mountains kept the land here bare and dry, erosion would began to slowly strip away the region's built-up layers of rock, creating a world of deep canyons and high plateaus, a place where the past would come back to life.

T H R E E

Modern Deserts and Ancient Forests

THERE ARE NO marked paths here in the canyon country that straddles the border between Canyonlands National Park and the Glen Canyon National Recreation Area. We find our way down to the canyon floor by clambering over ledges and steep slopes of sandstone, using cracks and holes weathered into the rock as handholds and clinging to the twisted trunks of pinyons and junipers for balance. Roughly one hundred years ago cowboys on horseback followed this same route down through the slickrock. They grazed their cattle in pockets of virgin grassland hidden among the maze of canyons below. For a time the deserts here offered a thin margin of profitability. Neither the profits nor the overgrazed islands of grass lasted for more than a few decades.

It is early August, and I am traveling into the canyon country of southeastern Utah with John Spence, a biologist with the Glen Canyon National Recreation Area, and his assistant Julie Zimmerman to study the relict forests of pine and fir that lie below the desert here hidden away in the shade of narrow side canyons and alcoves cut into the slickrock. Gary Cox, a backcountry ranger with the National Park

Service, travels along as our guide. The weather is hot. At nine o'clock in the morning the temperature has already begun to climb above ninety degrees.

In early spring the ground here had been covered by patches of snow that later gave way to flowers: purple stands of lupine, orange patches of mallow, yellow pools of wild marigold, and bright red tufts of Indian paintbrush. By late summer the ground is so dry and bare that it is impossible to imagine that it had ever been green and alive. The empty reach of the rock is broken only by scattered patches of sand covered with the black crusts of cryptogams and dry clumps of dead grass. Only the pinyons and junipers with their evergreen needles seem alive. They rise up out of the ground in twisted and contorted shapes, their roots reaching into cracks and seams in the rock.

The stream that flows through the canyon floor here is dry, its channel a bare sandy wash fringed by sagebrush and dry desert scrub. In places it opens wide enough to reveal islands of dry grass and small groves of cottonwoods. In places you can see cakes of dried mud and leaves left by spring floods when the floor of the wash was covered with moving water. In the dry summer heat the water has moved underground. It flows just below the surface, its path marked by inexplicable green ribbons of flowers and grass that rise up from the bare and seemingly dry surface of the sand. There has been no rain for more than four months. The summer thunderstorm season is several weeks old, but the clouds that appear each afternoon have brought nothing more than dry storms—high walls of clouds filled with wind and lightning that speed across the desert triggering brush fires and dust storms without leaving a single drop of rain behind.

Here and there as we head up the canyon we pass dry falls, high ledges of pink and mauve Kayenta sandstone that block the path of the stream. Flowing water has carved out hollows and holes and stained the rocks with ribbons of purple and black. Two hundred million years ago these purplish pink sandstones were deposited by shallow desert streams that often ran dry. Stopping to study the texture of the rocks, I can see the lines of ancient ripples and mud cracks and the strings of pebbles and stones left by ancient streams. Up above, layers of white and rust-colored rock, Navajo and Entrada sandstone, cover the canyon walls—the remains of ancient dunes that blew across the desert here some 190 million years before.

After a half-hour walk we turn up a narrow side canyon that leads back into the rocks. In a few yards its sandy floor gives way to a knee-high carpet of grass followed by an almost impassable tangle of low bushes and trees. Pools of water appear on the ground. A spring lies up ahead. Near the head of the canyon the trees and bushes suddenly fall away, to reveal a high, curving alcove cut into the surrounding walls of sandstone, a hidden amphitheater of pale red rock more than a hundred feet high. An apron of loose rock and sand leads up toward the base of the cliffs, canted at an angle of forty to fifty degrees. Thickets of dogwood and wild rosebushes cover their slopes. The trunks of towering evergreens rise up above them, a gallery of pillars and columns covered by the roof of the sky overhead. They are not dry pinyons and junipers like those found on the surface of the plateau above, but tall, spirelike stands of pon-derosa pine and Douglas fir sixty and eighty feet high. The largest have trunks more than three feet in diameter and may be more than a hundred years old. Beyond the dark branches of the trees, hanging gardens of monkey flower, maidenhair fern, and columbine cling to the smooth, curving surface of the rock.

After the long walk in the desert it seems cool and quiet in the shade of the alcove. A light breeze rustles through the trees. Boulders and stones on the ground here are covered by green layers of lichen and moss. As we unload our packs to start to work measuring and identifying the alcove's plants and trees, a flock of thirty to forty pinyon jays arrives with a raucous clatter. They land in an orderly row on a narrow ledge halfway up the curving walls of the alcove. Water percolates down through the porous layers of Navajo Sandstone above. Where it meets the more impermeable layers of the Kayenta below, the water magically reappears, seeping out of the rock like the beads of sweat that cover a cold glass of water. Quieting and settling themselves into an orderly row, the birds press their beaks to the sandstone, drinking from the thin sheen of water that covers the rock.

The plants here are unlike anything found in the surrounding desert. Flowering trees like dogwoods and redbuds are more like those one would expect to find in a hardwood forest in the East, while the pines and firs are more typical of what one would expect to find nearly a mile above on top of a high plateau or peak or perhaps farther north in Canada.

Plant communities on the Colorado Plateau are as much a product of evolution and change as its canyons and mesas. While it is relatively easy to visualize the rise and fall of a mountain or the steady wearing away of a canyon, we tend to see the forests and deserts that cover them as stable and constant. Traces of pollen from the floors of dry caves and lakes reveal that the climate here has changed dramatically over the past few thousand years. At the close of the Ice Ages some ten thousand years ago, the land here was both wetter and cooler. Forests of pine and fir similar to this one extended over much of the Colorado Plateau, reaching as far south as southern New Mexico. As the climate began to warm, desert plants like yucca, prickly pear, and Mormon tea moved in as the pines retreated northward and upward toward higher and cooler ground. A few, however, found enough shade and moisture to survive in protected alcoves and canyons like this one—islands of green amid the bare rock.

Today these alcove forests and hanging gardens are scattered throughout the canyon country like small green jewels. Not only are they found over a wide area, but populations of these isolated pine forests and hanging gardens are surprisingly alike, even those found at different elevations and with differing levels of sunlight, according to Spence. "I used to think it was just the luck of the draw which species were able to survive. Now I believe they were actually selected," he said. Varieties of Douglas fir, for example, while typically thought of in association with cool mountain forests, are actually highly variable, found everywhere from the Pacific Northwest to Mexico. Perhaps the onset of drought here tapped traits or abilities hidden within their genes. More drought resistant than the plants and trees that once surrounded them, they have been able to survive here for thousands of years. These hidden forests of pine and fir are living fossils that bring the past to life. In the floors of the canyons here you can see the past face-to-face

"These ancient forests are nothing more than long-term relics," said Kimball Harper, a biologist at Brigham Young University. "What we're looking at is a fragment of history: deep, startling, and unexpected." A few weeks after my trip into the canyons with Spence I had gone to talk with Harper, who has spent his career studying the flora and fauna of the Colorado Plateau and the neighboring Great Basin, at his office on the BYU campus in Provo. The plants and the trees of these canyon forests, he explained, may actually be a remnant of *Arcto-Tertiary flora*, an ancient belt of forests that extended around the Northern Hemisphere in a nearly continuous band more than 30 million years ago.

That idea of a common ancient forest was first proposed by the legendary Harvard biologist Asa Grey. A confidant and contemporary of Charles Darwin, Grey noticed that many of the plants and trees in northern Europe, Asia, and North America were surprisingly similar. When his plant collectors brought back samples from the islands of northern Japan, for example, he found that many of them were all-but-identical variants of the same plants and trees outside his office window in Cambridge, Massachusetts. Just as the migrations of continents have tied the world together, so too have the migrations of its plants and animals.

Here on the Colorado Plateau changing patterns of climate and terrain have sent species of plants and animals shuttling back and forth across the landscape, creating the complex but spare blanket of life that covers its canyons and plateaus today. Harper likes to think of ancient forests like the Arcto-Tertiary as only one of several patterns that have shaped the area, the warp from which its present-day flora and fauna have been woven. In addition to the possibly ancient origins of its alcove forests, the deserts and grasslands of the Colorado Plateau have complex histories of their own. Pinyon pines originated in the mountains of northern Mexico. Sagebrush first appeared in the steppes of central Asia and Siberia. It used to be that speculating on the possible origins of plants and ecosystems was a safe topic for conversation, Harper said. Now, however, the situation is rapidly changing. The same techniques of genetic analysis used to trace the evolution and origins of man in studies like the ambitious human genome project may also one day help biologists unravel the evolutionary history of plants and animals as well.

• • •

Just as the Colorado Plateau offers fertile ground for contemplating the evolution of the earth's continents and ocean basins, so too does it offer fertile ground for contemplating the origins of its plants and animals. The same barren landscape that makes the details of its rocks so easy to see also makes the details of its plants and animals easy to see. In places the ground is so sparse that you can pick out individual bushes and trees from several miles away—the twisted shape of pinyon or juniper on the face of a distant mesa or a solitary clump of saltbush or snakeweed on the floor of a vast desert valley. Plants here are not stacked up on top of one another, but often separated by a distance of several feet. There is none of the bewildering complexity and density of life that characterizes even the smallest woodlot in the Eastern United States or, for that matter, a coastal forest in northern California or the Pacific Northwest. The relationships between different species of plants and animals are easier to see. The land is spare and open. The world here has fewer moving parts.

But while life is spare here, it is also varied. In midsummer on top of places like the Henry Mountains in southern Utah you can stand on the edge of an alpine meadow of flowers and grass and look out over a burnt and barren landscape of dry grass and scrub in the deserts below. In between the high, green world of the peaks and the dry floor of the deserts are dry woodlands of pinyon and juniper; forests of ponderosa pine and Gambel Oak; and others still higher of aspen, spruce, and fir. The boundaries between these successive layers of desert and forest are often sharp and distinct. Rainfall and temperature change quickly with elevation. While the desert basins here typically receive less than ten inches of rain per year, the tops of high peaks and plateaus typically receive more than forty. While summer temperatures in the deserts often climb above one hundred degrees, up on top of the mountains and high plateaus they rarely climb above seventy. Populations of plants and animals here are often as layered as the rocks beneath them.

While the deep canyons below hide relict forces left over from the Ice Ages, at the same time the region's extreme topography has also created isolated populations of plants and animals where evolution is visibly at work. One of the best-known examples of this evolutionary isolation in the region is the Kaibab Squirrel, which inhabits the high pine forests of the Kaibab Plateau along the Grand Canyon's North Rim. A darker relative of the Abert or tassel-eared Squirrel that inhabits forests on the South Rim less than twenty miles away, the Kaibab Squirrel began to evolve only a few million years ago when the Colorado River began to cut through the Grand Canyon, effectively cutting these small animals from their relatives nearby. Similar less spectacular stories of evolution are seen in secluded populations of insects and grasses on top of the region's high peaks and plateaus. Much of the landscape here, however, is so young that such evolutionary variations are only just beginning to express themselves.

Down in the deserts below, the harsh demands of life have helped foster the development of other unique species of life as well: low clay hills with unstable, salt-laden soils and barren reaches of slickrock where the only purchase for plants comes in the forms of cracks and crevices. Life is resourceful and tenacious here. Walk through the slickrock deserts here after a summer thunderstorm and you will find pools of water in hollows and holes out on the barren surface of the sandstone filled with clouds of tiny shrimp and snails. Other forms of life here are as ancient as the rocks themselves. Here and there you can find patches of loose sand and soil covered with black crusts of cryptobiotic soil—mats of algae, lichen and moss that spring to life after the first sign of rain. Anchoring them to the ground are tiny cells of cyanobacteria or blue-green algae—modern-day relatives of the same simple cells of blue-green algae that appeared in the earth's ancient oceans more than 2,000 million years ago—the earliest known form of life on earth. Here on the Colorado Plateau the fabric of life reaches back from present and into the past—down through the roots of relict forests and cryptobiotic soils and into the fossils of ancient plants and animals that are now part of the rocks themselves. Quite often the plants and animals here have as much to say about the past as the layers of sandstone and shale beneath them. Locked up in their shapes, forms, and genes is a record of more than 2,500 million years of evolution and change.

F O U R

Those Who Came Before

T HE SKY IS blue and cloudless and heavy with heat. Out in the open desert the sun presses down on the ground like an oppressive weight. By late afternoon the sand will be so hot that it will be impossible to sit down. There has been no rain for four months, and the ground is very dry. Each footstep sends up a white puff of dust. Under a thin, low forest of pinyon and juniper a group of eight archaeologists from La Plata Archaeology, an archaeological consulting firm in Cortez, Colorado, is at work excavating the site of a prehistoric village. They work at a half a dozen or so excavations—long narrow trenches fifteen or twenty feet long and a collection of square-meter plots. They are not planning to excavate and map every square foot of the village, but only small cross sections—windows into the past that will offer a glimpse of what this seemingly nondescript site contains without destroying it completely should anyone equipped with promising new techniques or ideas wish to study it in the future.

There are no soaring pueblos of four and five stories bordered by plazas and giant kivas here. The traces of this simple, ancient village lie in the midst of an oil and gas field in northwestern New Mexico's San Juan Basin. It sits on top of a low, bare mesa overlooking a broad dry canyon. A few miles away the Animas River flows southward from Colorado toward Farmington, New Mexico, and the San Juan River. At regular intervals service trucks pass by the dirt road that borders the site, checking on pumps and pipelines farther out on the mesa. The whine and chug of oil pumps and gas compressors fill the hot, dry air with a steady drone.

Rather than academic research, their work here is part of a professional assignment. The site was randomly selected, part of a federally mandated study of archaeological resources in the area designed to offset the impacts of oil and gas development here and record at least a piece of the history it will inevitably alter and in some cases destroy. While oil and gas field hands tend to their pipelines and wells, archaeologists tend to the past, looking for signs of ancient peoples and cultures.

Progress is slow and tedious. In places tree roots lace the dry ground and have to be cut through with handsaws and axes, but aside from that the work is done strictly by hand, peeling back the dust and soil of several centuries with trowels, palette knives, and brushes. They do not have far to dig. Although artifacts suggest that the site here is more than two thousand years old, twenty centuries of time have managed to cover it with only a few inches of soil. The past lies just beneath the surface here. While some dig and sift through the thin, dry soil or cut through hardened layers of clay, others use a level and transit to carefully plot the location of features and artifacts. So far those details have been meager: small holes of blackened and hardened soil thought to mark the site of fire pits; a collection of manos and metates; circular or oval-shaped pans of hardened soil a few yards in diameter thought to mark the floor of a pithouse or shelter; a charred kernel of corn; and a small circular stone the size of a nickel with a hole through its center, possibly a bead.

During the first few days of work the site had seemed simple enough, said crew chief Leslie Hovezak. On the surface the ground had been littered with circles of fire-cracked rocks and stone tools from a Navajo hunting camp only a few hundred years old. A few inches below the surface, however, they found the remains of a cluster of Archaic pithouses, built by Archaic hunters and gatherers who lived in the Four Corners area from 5500 B.C. to roughly 100 A.D., several centuries before the rise of the Anasazi. They made no pottery and lived in simple pithouses, circular or oval-shaped pits dug a foot and a half or so into the ground and then surrounded by a low mud wall that was covered with a mound of sticks and mud.

In all probability these Archaic people were drawn to this spot by the same things that later drew the Navajo. The flat, sandy floor of the canyon below would have offered both land and water for small fields of corn and beans. Out on the edge of the mesa you can see up and down the canyon for more than a mile—an aid not only to hunting but perhaps security as well. In the late spring or early summer the heat would have been almost unbearable, but there is no need to assume that either these prehistoric people or the Navajos who followed them stayed here year-round.

Their simple way of life was not tied simply to the desert Southwest, but characterized much of prehistoric North America. Three to five thousand years ago, ancient peoples much like these inhabited the forests of Ohio and New York as well. But while this Archaic site lies only a few inches beneath the surface here in northwestern New Mexico, others of similar age and design in the greener and more fertile world beyond the Mississippi have all but disappeared—buried beneath several feet of soil, so reworked by burrowing insects and worms as to be completely unrecognizable. Here in the desert, however, it can take several thousand years to form a single inch of soil and in some cases it never forms at all. "Things last a long time here," Hovezak said. "That road out there will still be visible a thousand years from now," she added, nodding in the direction of the dirt road that leads past the edge of the site. The past runs no deeper here than anywhere else. It is simply easier to see.

These traces of early man are not unusual or unexpected. They are simply part of the landscape here. By some estimates there are more than 250,000 archaeological sites in the Four Corners area alone. Thousands of others lie scattered across the canyons and mesas of the Colorado Plateau and beyond—all the way from the Uinta Mountains to the Mogollon Rim. They range in size from scattered piles of potsherds to sprawling pueblos of several hundred rooms. Best known among these ancient peoples are the Anasazi, who left their pueblos and pictographs scattered across the desert Southwest. History, however, reaches both backward and forward in time from these ancient pueblo peoples, all the way from the hunting camps and kill sites of Paleolithic hunters who traveled through the landscape here during the Ice Ages to the abandoned mines and mill sites of the uranium boom that followed the Second World War.

The successive layers of the past here lie stacked up on top of one another like the layers of sandstone seen in the face of a cliff. Wander through the canyon that runs by the edge of the mesa here and you will find the ruins of Anasazi pueblos and granaries clinging to the sides of the cliffs. Hidden away in pockets and recesses in the canyon walls are panels of pictographs and petroglyphs left by the Ute and Navajo: multicolored images of horses and antelope overseen by galaxies of stars painted onto the surface of an overhanging ledge of sandstone. Elsewhere the rocks are etched with the names of passing Spanish sheepherders who ran their sheep in the deserts here: Lukas Martinez, Jose Luz. Separated by several hundred years of time and the perceptions of vastly different cultures, they drew their water from the same springs and streams and camped in the same canyons as those who lived here several thousand years before. History and culture on the Colorado Plateau are a product of steady evolution and change.

Ten thousand years ago roving bands of Paleolithic people hunted bison and woolly mammoths here. Descendants of the first people to migrate across the Bering land bridge to North America, they lived in caves and rock shelters and camped out on the land, following herds of game. Out of pieces of flint and chert they fashioned beautifully shaped spearpoints and blades, littering the ground with *debitage*, or flakes of worked stone. By some 5500 B.C. an Archaic culture based on hunting and gathering had developed—a more settled lifestyle based on the gathering of plants and seeds and the hunting of small game like rabbits, deer, and mice. In time they would develop a new agrarian way of life based on the cultivation of corn, beans, and squash—crops that had originally been discovered by other prehistoric peoples in Mexico—changes that would slowly transform these Archaic people into a people we know today as the Anasazi.

The first signs of this new way of life first appeared in the Four Corners area sometime between 100 B.C. and 400 A.D., as groups of the region's desert people began gathering together in small camps and villages where there were enough soil and water for farming. At first these early Anasazi lived in caves and pithouses just as their

Archaic ancestors had before them. While archaeological evidence suggests that they had already learned to cultivate corn and squash at this early date, they were still heavily dependent on wild foods like grass seeds and pinyon nuts. Known as the basketmaker people, they had no pottery but wove beautiful baskets from the stems and branches of plants they gathered in the desert.

For the next few hundred years changes were slow and incremental, as empty desert canyons gave way to small villages and crudely planted fields. Pottery appeared. Buildings made out of sticks and mud gave way to those made out of pecked and shaped blocks of rock and adobe.

Around 1100 A.D. the Anasazi suddenly burst into flower, creating one of the most sophisticated prehistoric civilizations in North America outside of the Mayan and Aztec worlds of Mexico. They did not stay near their fields on the canyon floors or the well-watered tops of mesas where life was easy and convenient, but moved up into the cliffs. At places like Mesa Verde and Canyon de Chelly they fashioned whole villages in the sheer walls of canyons, sprawling apartment houses four and five stories high that contained more than one hundred rooms. Plazas, kivas, and granaries were often several hundred feet above the canyon floor, reachable only by ladders and precarious handholds cut into the rock. Four hundred years before Columbus reached America, elegant towns and villages had been conjured out of the red rocks of the Colorado Plateau.

Their skill and artistry reached well beyond their cliffside villages. In their fields they developed an extensive system of irrigation canals and check dams for managing the scarce supply of rainfall. In their art they fashioned beautiful black-and-white pottery with elaborate geometric designs and decorated the sides of cliffs and large rocks with pictographs and petroglyphs. At places like Chaco Canyon in northern New Mexico they built signal towers and a system of roads that connected a network of more than one hundred villages and hamlets. Trade extended as far away as Mexico, bringing in exotic goods such as seashells and parrot feathers. It lasted no more than two hundred years.

By 1300 A.D. the Four Corners region and its picturesque cliff dwellings were all but abandoned. A number of theories have been proposed to explain why the area was deserted: drought, invading tribes from the north, even religious prophecy. There is also evidence that other more subtle and yet familiar factors may have played a role: an overgrown population coupled with poor farming techniques that exhausted the soil and left the high mesas stripped and bare. In the end the Anasazi may have been no more in tune with the world around them than our own present-day civilization.

Others would soon fill the vacant desert landscape left by the Anasazi. By the late 1400s bands of Navajo and Apache were moving into the area from their homelands in the forests of northern Canada while others like the Ute and Paiute were moving into the area from the deserts of southern Nevada and southeastern California. For each group the isolated landscape of the Colorado Plateau region would offer a chance to develop their own distinctive way of life. Each came to view the landscape in its own terms. The Ute would live off the land as roaming bands of hunters and gatherers right up until the late 1800s, when they were forced into reservations. The Navajo, in turn, would develop a pastoral lifestyle centered around herds of sheep. Moving seasonally with their flocks, they developed a way of life so closely tied to the land that they came to regard themselves not as immigrants from the north but as natives who emerged from right up out of the ground itself. The vast deserts and canyons of the Four Corners region became a holy land where gods and monsters once walked the earth, a landscape of beauty and power.

As for the Anasazi, they had not disappeared but simply moved south and east to form new communities like the Hopi and Zuni and dozens of even larger pueblos along the banks of the Rio Grande.

History here, however, did not stop with the departure of the Anasazi or the arrival of the Navajo and Ute. Less than fifty years after Columbus reached America, the Spanish Empire in the New World would begin pushing northward into the Southwest from Mexico. At first they viewed the deserts here as a barren wilderness to be claimed for the king of Spain and its natives as souls to be saved for Christ. For a time they dreamed of quick and easy wealth—rich mines of silver and gold and mythical lost kingdoms like the Seven Cities of Cíbola. Later they would be content to merely survive. In time they would come to view the land here in the same mystical light as the Pueblo and Navajo who had settled here before them. Their ties to it ran far deeper than the names they left on it: Santa Fe, Sangre de Cristo, San Juan. Martyred priests had given their lives here. Earth from the floor of a village church had the power to make cripples walk.

In spite of this deep reach of history, American pioneers and settlers passing through the area in the late 1800s still thought of the region as a vast, uncharted wilderness. Most took one look at the bare rock landscape and moved on. As the country moved west, the Southwest was left behind. Gold mines drew swarms of miners and prospectors to the mountains of Colorado and California, but the Colorado Plateau was an island of unexplored and impassable rock. Its geography would not be known with any certainty until well after the Civil War. Within its boundaries mapmakers and geologists would find the last unnamed mountain range in the lower forty-eight (the Henry Mountains) and the last unnamed river (the nearby Escalante).

For refugees like the Mormons who were fleeing religious persecution in the East, the seemingly barren deserts of the Colorado Plateau region and the neighboring Basin and Range country to the west became a place where they could build Zion, God's kingdom on earth. They came by wagon, foot, and horse from across the United States and Europe. They called themselves Latter-Day Saints. In the Old Testament Moses had led the Israelites out of captivity in Egypt to the Land of Canaan. For the Mormons, Brigham Young and his Quorum of the Twelve Apostles would lead them through the wilderness to the New Jerusalem. They did not settle helter-skelter on the landscape but came with a definite plan— laying out orderly towns with carefully planned streets and irrigation canals. Unlike most other western pioneers, the early Mormons were not loners or rugged individualists but part of a disciplined community. Like those who had come before them, they too saw the land as a chosen place. The hardships and isolation of desert life became a divine test of will and spirit. Names from the Bible and the *Book of Mormon* would be transplanted to the landscape: Moab, Manti, Ephraim, and Zion. When their crops were threatened by plagues of grasshoppers, God would send flocks of seagulls to save them. The routes of pioneers and pilgrims across the desert would become part of both legend and history.

The rest of the United States, however, would not reach the Colorado Plateau for nearly a century. While railroads and highways skirted the edge of the region, until after the Second World War much of it was traversed only by dirt roads and vague trails. Paved roads were almost unknown. Vast reaches of it were—and for that matter still are—accessible only on foot. As recently as the 1930s pack mules carried mail to the small town of Boulder, Utah, in the canyon country that lies along the flanks of the Aquarius Plateau near the Henry Mountains.

While the imprint of Native America runs no deeper than a string of picturesque place names over much of the United States, native peoples in the Southwest still live on land that has belonged to their ancestors for several centuries. Although cultures and traditions are changing fast, native peoples have been able to cling more tightly to their traditional ways of life than anywhere else in the United States outside of Alaska. Reservations in the region sprawl over several million acres, a collection of native states with their own laws, governments, and customs—independent nations within the larger framework of the United States whose history reaches back for several centuries. Most states in the region, by contrast, are less than a hundred years old and were governed as territories until the late 1800s—deemed too wild and unreliable for statehood. Most Anglo residents were often seen by outsiders as nothing more than a mix of religious fanatics, prospectors, cowboys, and whores—perceptions that were not, one might add, entirely without basis in fact. Utah, New Mexico, and Arizona were the nation's forty-sixth, forty-seventh, and forty-eighth states, respectively. Arizona did not achieve statehood until 1912.

That notion of frontier life has still not completely disappeared. Outside the boundaries of the region's Indian reservations, tens of millions of acres are preserved as public land in a collection of national parks, national forests, and Bureau of Land Management lands. With a modest list of regulations, much of it is as open to the public for travel and exploration as it was several hundred years before—a broad reach of open, accessible, and largely unpopulated land.

In spite of the vast reach of the land here, one must keep several conflicting ideas in mind to understand the present-day landscape. While the Colorado Plateau region is largely uninhabited, its parklands—Grand Canyon, Zion, Bryce, Lake Powell, Canyonlands, Arches, and Mesa Verde—are filled to overflowing. Today the same rugged landscape of deep canyons and high plateaus that kept all but the most desperate pioneers and settlers at bay draws tourists by the thousands. While the traditional cultures of the region's native and Hispanic peoples seem more alive here than anywhere else in the United States, they are disappearing quickly as its people are forced to cope with the inescapable poverty of the land and the realities of making a living in the more competitive and more modern world that surrounds them. While the land here seems wild and untouched, much of it is far from unused. Dams have transformed its wild rivers into thin winding lakes, strip mines gather coal for giant power plants, herds of cattle and sheep wander through desert valleys on vast reaches of open range miles from the nearest town or ranch.

Today the culture of the region is a mixture of Indian, Spanish, and Anglo life. Whatever their reasons for coming here, each of these peoples has had to come to terms with the landscape: its ruggedness and dryness. Their stories and histories are often as intriguing as those of the land itself. Just as the present-day landscape has been cut from the built-up layers of the past, so too has the present-day culture of the region been cut from the layered histories of its people. The people of the Colorado Plateau today are the product of more than five centuries of interaction and change. Nowhere else in the United States do so many different peoples converge. Nowhere else in the United States is the history of the land and its people so closely tied together.

BOOK TWO

~~~~~~

North

THE PHYSICAL LANDSCAPE is baffling in its ability to transcend whatever we would make of it. It is as subtle in its expression as turns of the mind, and larger than our grasp; and yet it is still knowable. The mind, full of curiosity and analysis, disassembles a landscape and then reassembles the pieces—the nod of a flower, the color of the night sky, the murmur of an animal—trying to fathom its geography. At the same time the mind is trying to find its place within the land, to discover a way to dispel its own sense of estrangement.

BARRY LOPEZ
Arctic Dreams

FIVE

The View from Split Mountain

THE RIVER SLICES through Split Mountain as if it wasn't even there. Low, flat ground lies to either side, but the Green River cuts a curving arc right through its highest point, exposing its layered core to plain view. Canyon walls are covered in part with the same Mesozoic rocks found farther south at places like Zion and Canyonlands National Parks: brightly colored layers of Moenkopi, Chinle, and Navajo, a collection of sandstones and shales. They do not lie flat, but are tilted almost vertical, following the mountain's profile. The mountain's details are as puzzling as the course of the river itself. In fact, Split Mountain is not really a mountain at all but a curving wall of rock some five miles long. The bent and tilted layers of rock within it seem to have risen up out of the desert like a wave only to break along the edge of the higher plateaus that lie beyond it, the shoreline of an unseen dry sea.

Down in the canyon where the river tumbles over boulders and stones, you can see the white sheen of rapids. Green, parklike rows of cottonwood trees line the river's edge along the wider reaches of the canyon floor. Upstream lies Echo Park and the junction of the Yampa River flowing west from northern Colorado. Above that lies the Canyon of Lodore, named by John Wesley Powell and his party of largely amateur explorers who floated through the canyons here in the late 1800s. Their historic trip through the region began a few dozen miles upstream at Green River, Wyoming, and would bring the Colorado Plateau and its isolated canyons to worldwide attention. From Wyoming they would follow the Green River to its junction with the Grand, as the upper reaches of the present-day Colorado River were known in the mid-1800s, near Moab, Utah. From there they would follow the Colorado southwest through the Glen Canyon and finally westward into the Grand Canyon itself. Originally the expedition had been planned as a careful scientific survey of the canyons of the Green and Colorado, but by the time Powell and his men had reached the final wall of the Hurricane Cliffs, where the river flows out of the Grand Canyon and into the low, hot deserts that straddle the Arizona-California border, they would be fighting for their lives.

Here in the north, Split Mountain is a kind of final wall as well. Beyond it the river empties out into the flat, barren deserts of the Uinta Valley. No longer confined by high plateaus or canyon walls, it becomes, for a brief time, a broad, sandy stream, flowing southwest toward the Tavaputs Plateau and Desolation Canyon, where the land rises up to embrace the river again.

South of the mountain the land is wide and open—a khaki-colored landscape of tough desert plants no higher than your knee that stretches southward as far as the eye can see. Beneath the clear, late-summer sky it seems burnt and dry—a sea of sagebrush, cheatgrass, and greasewood bordered only by the thin, low lines of distant plateaus.

To the north and west, however, the view is green and mountainous, a landscape of high plateaus covered with forests of ponderosa pine and Douglas fir, fringed with woodlands of pinyon and juniper. Looking west across the thin, watery line of the Green River, I can see the soft, rounded shapes of the Uinta Mountains that define the northern edge of the Colorado Plateau. Late-afternoon rainclouds float above them, part of the daily buildup of rain that marks the summer thunderstorm season in the Intermountain West.

The Uintas are a dividing line of sorts, separating the plateau and canyon country of the Colorado Plateau from the sprawling reach of the northern Rocky Mountains. After passing through Colorado in a fairly narrow band, the Rockies fan out in a broad belt of high peaks that stretches all the way from central Wyoming to Idaho. Broader to the north, they seem to almost wrap around the Colorado Plateau, bordering not only its eastern edge but its northern edge as well.

The Uintas are the only mountain range in the Intermountain West that runs east–west. Mountain ranges within both the Rocky Mountains and the Basin and Range country that lies to either side trend roughly north–south. The Uintas have a well-worn profile that gives them the appearance of great age, and that appearance is not deceiving. Rocks from the mountains' core are nearly as old as those found on the floor of the Grand Canyon several hundred miles to the south. Ancient patterns of faults and folds more than one billion years old have left behind a zone of weakness in the earth's crust that later episodes of mountain building have exploited to raise these peaks up at odd and incongruous angles to the terrain around them. Their most recent rise began some 50 to 70 million years ago.

Split Mountain and the plateaus and canyons that surround it are all part of Dinosaur National Monument, several hundred thousand acres of high desert parkland spread across Utah and Colorado along the course of the Green and Yampa rivers. In the 1950s the canyons that cut through the park were slated for dams that would have turned them into narrow lakes. Opposed to the idea of dams in a federal park, a coalition of conservation and environmental groups banded together to fight the Bureau of Reclamation's plans and won. It was the beginning of the end of the American love affair with large dams and federally subsidized water projects. For a brief time the park was front-page news.

Today, however, while parks to the south like Grand Canyon, Arches, and Zion are overrun with tourists, Dinosaur seems almost forgotten. While summer brings a steady stream of tourists, few venture much further than the fossil quarry that lies on the flanks of Split Mountain, where paleontologists have found some of the most spectacular collections of dinosaur bones in the world—the skeletons of giant beasts like the stegosaurus and brontosaurus that roamed through the swamps and lakes that covered the dry landscape here some 140 million years ago. The rest of the park remains relatively untraveled.

Farther north inside the park, at places like Harpers Corner above the junction of the Green and Yampa, you can look down into the depths of Echo Park and Whirlpool Canyon, where the rivers wind through multicolored layers of Paleozoic limestone and sandstone of the same age and appearance as those found in the upper reaches of the Grand Canyon.

Here too it is not just erosion that has brought these ancient rocks into view but a series of nearly vertical faults that have broken the ground into blocks that have been alternately raised and lowered like stairsteps. In places on the canyon walls you can see the traces of those faults quite clearly. In places movement has caused the rock layers to curl and bend, offsetting level layers of rock by several dozen feet, as if they have been bent by the force of movement and slip.

Out on the thin promontory of rock that stretches out from the cliffs you can stand on top of rocks that were deposited on the floor of an ancient sea more than 500 million years ago and see the fossils of stromatolites—thick mats of blue-green algae that grew as dome- and pillow-shaped mounds on the seafloor. A few feet away in the forest of pinyon and juniper that borders the cliffs, you can see the modern relatives of those fossilized algal mats alive and well in the black patches of cryptobiotic soil that cover the desert floor—tangled, living mats of algae, lichen, moss, and bacteria that thrive in the high deserts here. Scientists believe that simple crusts of plants and bacteria like these may well have been the first forms of life on earth— pioneers that prepared the way for the rise of more sophisticated forms of life. Here in the desert several hundred million years later, they perform similar pioneering functions as well—an anchor for life that binds the loose desert soils together and fixes valuable nutrients like nitrogen out of the high, thin air. Patches of these cryptogamic soils can be found almost anywhere in the desert reaches of the Colorado Plateau. Simple and tenacious, they offer a glimpse of life at its most basic level.

Signs of early man are here as well. Nearby canyons are filled with pictographs and petroglyphs of prehistoric pueblo peoples like the Fremont, who lived here more than a thousand years before. Elsewhere are even older signs of human life: the hunting camps and kill sites of Paleolithic hunters who stalked mammoths and mastodons here more than nine thousand years ago. If the deeply buried rocks exposed in the canyon walls are the stable base of the landscape here, these Paleolithic peoples are the stable base or core of the region's native peoples. Descendants of the first roving bands of hunters to reach the New World from Asia, they carried their hunting and gathering lifestyle throughout North and South America. Out of that simple way of life a host of cultures and peoples would evolve—not only the Fremont and Anasazi here in the Southwest but people as diverse as the Mayans and Aztecs of Central America and the Seminole and the Sioux of the eastern United States.

Here in the Four Corners area that simple way of life would never completely disappear. Several centuries after the canyons and deserts here were abandoned by the Anasazi, their place would be taken by roving bands of hunters and gatherers like the Ute and Paiute who migrated into the area from the deserts of southern Nevada and southeastern California: newcomers whose nomadic way of life was almost indistinguishable from that of the Paleolithic hunters who had settled here several thousand years before them. Although the horse would dramatically increase both their range and their wealth, the Ute would continue their hunting and gathering lifestyle right up until the late 1800s, when they were herded off to reservations in Colorado and Utah.

Two of those, the Uinta and Ouray reservations, lie just south of here along the banks of the Green River, sprawling across the flat, open reaches of the Uinta Valley and the rugged terrain that fringes Desolation Canyon on the East Tavaputs Plateau. Others lie in Colorado both south of Durango along the New Mexico border and over toward the Four Corners in the low deserts that surround the southern flanks of Mesa Verde. If the modern-day pueblo people can be seen as a living continuation of the ancient culture of the Anasazi, others like the Ute can be seen as the living continuation of the even older culture that preceded them—the very roots from which the more sophisticated world of the Anasazi would spring.

The broad flat deserts of the Uinta Valley are a sharp contrast to the maze of deep canyons and intricately carved slickrock deserts that dominate the landscape farther south. Even here, however, the land is not featureless. Lines of low dry hills conceal small hidden canyons whose floors are laced with improbable meadows of bright green grass. Out amid the rolling surface of the desert you can find herds of antelope and half-wild horses hidden in the protection of a narrow draw or a shallow swale. As the land rises up toward the Tavaputs Plateau and the Book Cliffs farther south, desert grasslands slowly give way to forests of pinyon and juniper. Before the land here was uplifted and the rivers began to carve its surface into a series of deep canyons and high plateaus, much of the Colorado Plateau looked like this: a flat and almost featureless plain. Here in the high deserts of northern Utah and Colorado you can see how both the land and its people took shape. You can get down to basics. In time the rivers that now wind through this open terrain will strip away the dull blanket of rocks that covers it, just as they have farther south in the maze of canyons that lies near the plateau's center. Here in the north both the future and the past lie buried beneath the present.

S I X

Ancient Mountains, Ancient Borders

T HE PLANE LIFTED slowly off the runway, the transition from
ground to air so smooth it was barely noticeable. As we rose up
out of the flat, desert plain north of Moab, the landscape below
began to take on shape and form. Ahead, the high wall of the Book
Cliffs came into view. Flat, even layers of tan and brown rock run
across their face, broken by thin lines of black and gray—the sandstones and shales of the Mesaverde Group that were deposited on
the edge of a retreating ancient sea that covered the landscape here
some 70 to 80 million years ago. Over the past few million years the
cliffs have retreated northward in a curving wall, peeled away like
old layers of varnish to reveal the details of the older landscape
below.

As we turned west to head toward the San Rafael Swell and the
high plateaus of the Wasatch Front, I could see domes of red and
white rock rising up out of the ground to the south, the slickrock
deserts of Arches and Canyonlands National Parks. Here to the
north, however, the rocks below are not red but purple, green, and
white—a collection of shales and sandstones known as the

Morrison Formation, left by a series of lakes and streams that covered the ground here some 160 million years ago. Here and there shelves and ledges of sandstone protrude out of the desert like ribs. Farther north at Dinosaur National Monument layers of rock from this same formation are laced with the fossils of dinosaurs, crocodiles, and turtles. Here the paths of ancient streams are collecting points for uranium, and the desert below is laced with a spider's web of dirt roads and the pits of small, abandoned mines—part of the uranium boom that brought a small but determined army of prospectors and miners to the Colorado Plateau in the years that followed the Second World War.

It was early morning, and the sun was only a few degrees off the horizon. Long, low rays of light shot across the ground, illuminating the cobbled paths of ancient streams. Erosion has turned the landscape here inside out. Harder and more resistant that the softer shales and sandstones around them, the pebbles and cobbles of ancient streambeds form a protective cap for the softer rocks below. Instead of lying at the bottom of a shallow wash, they rise above the low, level surface of the ground in winding ridges of gravel. In the sharp light and shadow of early morning they seem to stand out in bas-relief, running across the surface of the desert like giant snakes.

It was the first week of April, and I was flying with Bill Dickinson, president of the Geological Society of America, in a rented Cessna. On board was Bill's wife, Jackie, and our pilot, Lavar Wells with Redtail Aviation in Moab, Utah. Wells grew up near Hanksville, Utah, and has been flying over the Colorado Plateau for more than forty years. There were no paved roads into Hanksville when Wells was growing up, and his father found flying preferable to driving when it came to making long trips. While the region was remote, it also was studded with makeshift landing strips. At an age when most young people are still waiting for their first driver's liscense, Wells had already learned to fly. As a professional pilot, Wells has more than ten thousand hours of flight time, much of it gained right here flying over the deserts and canyons carrying everything from supplies for mining camps to parties of rafters bound for trips down the Green or Colorado.

Our own trip was the start of three days of flying around the plateau, aerial field trips in effect, planned to give us a new perspective of the patterns of faulting, folding, and erosion that have shaped the landscape here. Our route that day would take us in a semicircular arc around the northern rim of the Colorado Plateau—due west across the San Rafael Swell to the Wasatch Plateau and then eastward across the southern edge of the Uinta Valley to follow the Green River south through Desolation Canyon and back across the San Rafael Desert to Moab.

As we headed west across the desert, the tilted rock walls of the San Rafael Swell came into view. Created by slip along the surface of a nearly vertical fault that has bent the ground above it into a broad bow or arch, it runs across the desert for more than 40 miles like the track of a giant mole. Along its outer edge, resistant layers of rock have been broken and worn into flatirons, tilted wedges of Navajo Sandstone that rise up out of the ground like teeth or the plates of armor that line the back of a stegosaurus. Behind them are multicolored layers of Kayenta, Wingate, Chinle and Moenkopi, the same layers of Mesozoic rocks that line the walls of Zion Canyon farther south. Heading toward the center of the swell, the rocks become progressively older. The tilted rocks walls of the swell were once part of a high, domelike fold. Today only its flanks remain. Its crest has been completely eroded away—exposing the older Paleozoic rocks below. Rocks near the center of the swell are white, the same Kaibab Limestone that makes up the rim of the Grand Canyon several hundred miles to the south. Farther west the same sequence is repeated in reverse as the eroded flanks of the swell reappear and then dive beneath a protective cover of Mancos Shale and the high, table-like rise of the Wasatch Plateau beyond.

The San Rafael Swell took shape some 50 to 70 million years ago, part of the Laramide orogeny that built up the Rocky Mountains some 300 miles to the east. In places its crest rises up more than 500 feet above the surrounding desert, but the high rise of the rocks does not seem to bother the San Rafael River at all. It cuts right through the heart of the swell in a deep canyon, the gap between its curving walls so narrow that in places cowboys on horseback are said to have actually leaped across it.

There are several other broad folds like this one scattered across the face of the Colorado Plateau—the Waterpocket Fold, the Circle Cliffs Upwarp, the Comb Ridge Anticline. Early Mormon pioneers called these tilted walls of rock "reefs" because they were barriers to travel. In the late 1800s when bands of pioneers began to push south-east from Salt Lake City to settle the red rock deserts of the plateau region, they found these high, tilted walls of rock almost as insur-mountable as the region's deep canyons. Traveling by air, a thousand feet above the ground at one hundred miles per hour, they look almost unreal—an illustration of faults and folds from a geological textbook, suddenly come to life.

"This is what it's all about," Dickinson said with a smile as he looked out the window. "This is what you're trying to do all the time in your head. You want to see how things fit together. When you're looking at an outcrop on the ground or studying a map, your mind is constantly trying to put all the little pieces together. You don't have to do that up here. It's all right in front of you."

Dickinson was born and raised in Tennessee. In 1946, when he was sixteen years old, his family moved to California, traveling west on Highway 66. He saw his first butte outside of Tucumcari, New Mexico. They passed through the Painted Desert and stopped to look at its "forest" of petrified wood. Although Dickinson did not know it, a few years later he would be back on the Colorado Plateau.

He was sitting on the beach in Santa Barbara when, as Dickinson recalled it, "A crazy man drove up in a DUKW, an amphibious army assault vehicle left over from the Second World War. It was painted yellow and blue with 'Explorers Camp' written on it." Dickinson wandered over and struck up a conversation. The man with the DUKW was Kenny Ross, and he ran something known as the Explorers Camp in the Four Corners area, a kind of Outward Bound program for teenagers and college students several decades before its time. Dickinson had grown up on a farm in Tennessee and knew how to operate trucks and heavy equipment. He also knew how to pack horses and had spent time canoeing on nearby rivers and streams—qualities that all seemed to lend them-selves to Ross's own line of work. They hadn't talked very long before Ross had offered him a summer job as a camp counselor. For his last years of high school and his first few years of college, Dickinson spent his summers working on the rivers and in the back-country of the Four Corners area as a guide. They ran the San Juan and Colorado rivers and explored the maze of canyons that led into Cataract Canyon from the flanks of the Abajos—Dark, Gypsum, and Elk—on ten- and fourteen-day backpacking trips. They traveled for weeks at a time without seeing a soul. Rafting was more of the same. "We never saw anyone on river trips on the San Juan. We never expected to and never did," he said. A few times they started in Bluff, Utah, and ran the San Juan to the river's junction with the Colorado near Mexican Hat, and then followed the Colorado southward through the Glen Canyon all the way to Lee's Ferry. Today everything is different. "In the past fifteen years I've never taken a trip and not seen a boating party on the river. Four or five parties launch every day and you're constantly passing and being passed by one another," he said.

In the 1940s and early 1950s most of the Four Corners region was traversed only by dirt roads that followed the paths of dry washes and streams. There was no pavement west of Shiprock in New Mexico or north of Cameron, Arizona. Driving from Cortez to Kayenta—a two-hour trip today—could take several hours or even several days—busting through sand traps in the desert and skirting patches of quicksand in the canyons. "You could spend all day traveling to places that are only a short morning's drive away today," Dickinson said. "If the streams in the washes came up you had to wait for them to go back down. You didn't have any choice. If it was raining, forget it. You might be washed out the other end."

Ross had been an archaeologist at Mesa Verde, a photographer, and filmmaker and had a strong scientific bent as well. As Dickinson and the group's other guides and counselors worked, he taught them about the region's geology and natural history. Walking and rafting through the region's canyons and deserts, they learned the names of the Colorado Plateau's rock units and developed an eye for identifying them in the field. While Dickinson found geology fascinating, he still thought of it as more of a recreational activity than a profession. "I thought you had to earn your living doing something more prosaic," he said. In college at Stanford he dutifully continued majoring in engineering. He took his first class in geology as a junior and was quickly taken in, but it was too late to change his major. After finishing his undergraduate degree, he started graduate school at Stanford—this time in geology, continuing on to eventually teach as a full professor before going on to the University of Arizona, where he became department head.

His career took shape during the plate tectonics revolution that would reshape the science of geology, and Dickinson would spend much of his career helping unravel its finer details. Although his interest in geology first took shape here in the Colorado Plateau region, he spent most of his professional career working around the margins of the Pacific in places like California, Oregon, Fiji, and Japan. "I got interested in the Pacific Coast because of its complexity," he said. "For most of my professional career I steered clear of places as little deformed as the Colorado Plateau."

While the plateau's vast exposures of rock are a geologic gold mine for those interested in the intricacies of sedimentary geology—the fine details of the sands and silts left by ancient rivers, dunes and seas, the plateau has little to say about the rise and fall of mountain ranges or the movements of continents and ocean basins. For those like Dickinson whose interests are more structural—studying the effects of faulting, folding and uplift—the flat-lying rocks of the Colorado Plateau have only a limited appeal. "It's only interesting in a regional sense," he said of the area's structural history, "because that's where it gets complex enough to be really satisfying." And over the past few years that regional sense of geologic history is something Dickinson has found increasingly intriguing. The Colorado Plateau is a geologic riddle, an island of stable and solid ground surrounded by a mountainous sea of disrupted and chaotic rocks. In the past, Dickinson said, the Colorado Plateau may have been far larger than it is today. Over the past few hundred million years repeated episodes of faulting, folding and mountain building have carved off large pieces of terrain, leaving the plateau behind like the random pieces of dough left by a cookie cutter. While the canyons that slice through the landscape here are perhaps no more than five million years old, the features that define the borders of the Colorado Plateau today were often marked out by events that took place hundreds of millions of years ago.

As we headed west toward the high plateaus of the Wasatch Front, we passed over low mesas covered with scattered stands of pinyon and juniper and meadows of dry grass where herds of wild horses ran across the ground. Flying along the base of the Wasatch Plateau we passed over a string of tiny Mormon towns—Fairfield, Castledale and Orangeville, connected by the thin lines of roads and irrigation canals.

Climbing over the 11,000-foot-high crest of the plateau, we rocked and bounced in the sky as the plane was buffeted by turbulence. Down below the snow still lay in deep drifts amid forests of aspen, spruce and fir. Heading north along the broad, high spine of the plateau, we were between worlds. To the east was the Colorado Plateau—a world of flat-topped mesas, deep canyons and broad deserts stretching all the way to the Rockies. To the west was the Basin and Range—a world of long, high mountain ranges and dry, narrow valleys that run all the way to the Sierras. Looking across its successive ridges of high peaks, you could see all the way to Nevada and the snow-covered peaks of the Snake Range.

Streams flow down the high peaks of the Basin and Range here, only to evaporate in the land-locked valleys below. Fifteen to twenty thousand years ago during the Ice Ages those land-locked valley floors were covered by freshwater lakes. Today they have all but evaporated, leaving behind stark white plains of salt like the Bonneville Salt Flats that lie due west of Salt Lake City and shallow playa lakes. Today the saline waters of Utah Lake and the Great Salt Lake farther north are all that remains. Once filled with fresh water, evaporation has concentrated their load of dissolved salts and minerals. Today their waters are saltier than those of the Pacific more than 800 miles away.

The line of this boundary between worlds is not random. It has been a border several times in the last 500 million years—a line of demarcation. The same late-Paleozoic rocks that line the walls of the Grand Canyon and reach eastward beneath the Rockies run westward toward the Great Basin as well. Beyond the line of the high plateaus here, however, these Paleozoic rocks are almost entirely marine and rapidly grow in thickness. Perhaps, some geologists have suggested, the shift from shallow water to deep water rocks marks the trend of a hingeline, the edge of an ancient continental shelf that separated the shallow water near shore from the open ocean beyond. Whatever its origins, this ancient boundary would become a focal point for activity, used over and over again by the forces that shaped the western edge of the continent.

Some one hundred and fifty miles to the west in Nevada is the edge of the continent's craton—the edge of the stable platform whose 2,000-million-year-old rocks are exposed on the floor of the Grand Canyon to the south. West of this ancient edge the land is all Suspect Terrane—belts of rock and fragments of continents and island arcs that have drifted through the Pacific for dozens or perhaps even thousands of miles before colliding with the edge of the continent to build up California and Nevada.

Traces of those collisions are close at hand. The Sevier Mountains just off the western flank of the Wasatch Plateau outside our plane's windows were created by an episode of ancient mountain building known as the Sevier orogeny that took place some 80 to 100 million years ago. Driven by plate movements and collisions farther west, compression caused by the Sevier orogeny shortened what is now the Basin and Range country of Nevada and Utah by some 40 to 60 miles. In places the ground collapsed like an accordion, but here near the edge of what is now the Colorado Plateau, it thrust layers of rock up over the top of one another, stacking them up like bricks. The Sevier Mountains mark the leading edge of that ancient thrust sheet. When they appeared, however, the high plateaus of the Wasatch front were nowhere to be found. Rivers and streams flowing off the flanks of these thrust-up mountains carried sand and gravel from their slopes eastward for hundreds of miles all the way to the Rockies. The crest of those newly created mountains, however, coincided almost perfectly with the ancient boundary between shallow shoreline and deep sea that had existed some 500 million years before. Fifty million years after the Seviers had appeared, the once-compressed ground behind them would begin to stretch and thin, breaking the earth's crust here into blocks that tilted and sank like blocks of ice in a half-frozen river, creating the rafts of high mountains that stretch across the region today. That disturbance, however, would reach no further into the Colorado Plateau than the episodes of thrusting that had preceded it. It was almost as if there was something different or unique here, some impenetrable ancient boundary that could not be crossed.

Similar ancient boundaries define the northern reaches of the Colorado Plateau as well. As we neared Provo and the south shore of Utah Lake we turned east and headed toward the Green River, crossing over the tangle of mountains and plateaus that skirts the broad flat reach of the Uinta Basin. To the north the high, snow-covered peaks of the Uinta Mountains ran due east along the edge of the horizon. They are unlike any other mountain range in the Intermountain West. While the high peaks of the Rockies and the Basin and Range that lie to either side run north—south, the Uintas run east—west. Their structure is peculiar as well. Thrust faults border both the northern and southern flanks of the mountains and seem to converge beneath them as well—almost cutting the high block of the peaks off from the rocks below. Even here from some fifty miles away you can see how the sandstones and shales in the Uinta Valley seem to dive beneath the mountains. Layers of rock do not tilt upward over the mountains, but seem to dive down into the earth—as if the high peaks behind them had been simply dropped out of the sky, bending the rocks beneath them into a broad trough.

Rocks in the core of the Uinta Mountains are more than 1,000 million years old. While mountain-building and metamorphism were reshaping the ancient schists that line the floor of the Grand Canyon, the rocks that lie within the core of the Uinta Mountains seem to have been part of a deep basin or trough—possibly, some geologists have suggested, part of a failed ancient rift zone. While that rift zone ultimately disappeared, it left a zone of weakness behind that later episodes of mountain-building were able to use to raise these mountains up to odd and incongruous angles to the ter-rain around them to define the northern edge of the Colorado Plateau. Like the Rockies to the east and the San Rafael Swell to the south, they were uplifted as part of the Laramide orogeny that took place some 50 to 70 million years ago.

Today these rocks from the floor of an ancient basin have been uplifted into mountains nearly two miles high. Like waves at sea that are followed by troughs, however, mountains on land are often accompanied by the appearance of basins and the Uintas are no exception. Deep basins lie to either side of the mountains, collecting points for the sand and silt that erodes off their flanks. While rocks in the core of the mountains are more than 1,000 million years old, those that fill the Uinta Basin that lies here to the south are less than 100 million years old—a collection of sandstones and oil-rich shales two to three times as deep as the mountains are high. The borders of this filled-in basin roughly coincide with those of the Uinta Valley above.

The southern edge of the Uinta Valley is defined by the high rise of the Tavaputs Plateaus, their tops covered with forests of ponderosa pine and gambel's oak, rising several thousand feet above the deserts below. The Green River cuts right through this high rise of the ground in the deep reach of Desolation Canyon, between the East and West Tavaputs Plateaus. We followed the twisting path of the river southward through the high wall of the Roan Cliffs on the edge of the plateaus. In the curving walls of the canyon you could see the same layers of rock that fill the Uinta Basin to the north laid out in staggered rows, huge amphitheaters of finely layered red- and tan-colored rocks.

In the south as the river emerged from the Book Cliffs and flowed out into the broad, flat reach of the San Rafael Desert, we took a curving route east toward the La Sal Mountains outside of Moab. Far to the east, the western edge of the Rockies was just barely visible, marked by the high reach of the San Juan Mountains some 100 miles away, their snow-covered peaks looking almost like clouds on the edge of the horizon. Between them and the desert below was the long, mound-like rise of the Uncompahgre Plateau. Like the Uintas farther north and the Rockies farther east, the Uncompahgre Plateau was uplifted as part of the Laramide orogeny. Like the high plateaus of the Wasatch Front to the west and the Uintas to the north, the Uncompahgre Plateau has an ancient history of its own as well. Some 250 to 350 million years ago the Uncompahgre was uplifted as part of the Ouachita orogeny that built up a range of ancient mountains here known as the Ancestral Rockies. Unlike the Laramide that gave rise to the modern Rockies, the Ouachita orogeny that built up the Ancestral

Rockies before them, was not related to collisions and plate movements taking place farther west, but to those taking place farther south. Roughly 300 million years ago, pieces of Central America collided with North America, roughly along the line of what is now the Gulf of Mexico. The collision was part of the plate movements that would eventually link the continents together into a single land mass by the close of the Paleozoic some 245 million years ago to create the supercontinent of Pangea. The force of the collision built up mountains not only here in Colorado where the Uncompahgre Plateau and the Rockies now stand, but to the southeast in Oklahoma and Arkansas as well. While those ancient mountains would be completely eroded away in Arkansas and Oklahoma, farther west they would rise again in the Laramide orogeny nearly 250 million years later and build up an entirely new range of peaks.

The rise of the Ancestral Rockies, however, left far more than just planes of weakness. As we neared the La Sals, the rolling slickrock terrain of Arches National Park passed beneath us—a multicolored layercake of Navajo, Entrada and Carmel sandstones. From above the rocks seemed blistered and cracked. The ground was not flat like the surface of the San Rafael Desert farther north, but broken by broad domes and bubbles of rocks—their tops scored with a gridwork of joints and seams.

The rolling surface of the rock here was not caused by erosion that wore the rocks away from above, but by domes of salt rising up from below. A mile beneath the surface are beds of unstable salt several thousand feet thick formed by the evaporation on the floor of an ancient sea that covered the landscape here more than 300 million years ago. Just as the rise of the Uinta Mountains to the north was accompanied by the appearance of the adjoining Uinta Basin, the rise of the Ancestral Rockies and the Uncompahgre Plateau was accompanied by the appearance of a broad downwarp in the earth's crust known as the Paradox Basin. It stretched for nearly 200 miles from Moab to Durango, a shallow sea where the eroding sediments of neighboring highlands accumulated. As this ancient sea slowly filled with debris, it was gradually transformed into a vast pool of shallow, saline water—perhaps not unlike the Red Sea that lies between Africa and the Arabian Peninsula today.

Like water in a pan that boils completely away, evaporating sea-water left crusts of salt and minerals behind. Several million years of steady evaporation built up deposits of salt and gypsum more than a mile thick—known today as the Paradox Formation. Over the next few hundred million years a succession of ancient seas and sand dunes would bury it with several thousand feet of rock, setting the stage for the later rise of the salt. Less dense and more fluid than the rocks around it, giant domes of this buried salt began to rise up through the overlying layers of rock like bubbles of hot tar. Where they rose to the surface they caused the land above them to dome and swell, creating the peculiar hummocky terrain that covers much of Arches and Canyonlands National Parks today.

In places near the surface, however, groundwater slowly washed these water-soluble domes of salt away, leaving behind hollow caverns and shafts. The collapse of these salt caverns has left a string of narrow valleys scattered across the high deserts that surround the La Sal Mountains. Spanish Valley that runs south from Moab toward Monticello was formed by the collapse of one of these ancient salt domes. So too were Paradox and Gypsum valleys farther east in Colorado. They seem to appear in the desert without rhyme or reason. Here on the Colorado Plateau the past has an odd way of reasserting itself.

S E V E N

Potholes and Cryptogams

It began as a light drizzle shortly after midday. By late afternoon the rain was so heavy that it seemed almost tropical, pouring out of the sky like water out of a spigot. It flowed across the bare rock in sheets, like water running down a city street after a heavy rain. Out along the edge of the canyon it tumbled over ledges and cliffs turning dry walls of rock into waterfalls. Down below rivers of red and brown water flowed through the side canyons and across the floors of once-dry washes. Several days later I would learn that the storm had dumped more than three inches of rain in a single hour. It seemed incredible that such a dry landscape could be soaked with so much water.

I was out on the White Rim, the broad bench of rock that fringes the canyons of the Green and Colorado as they wind through Canyonlands National Park in southeastern Utah. The rim lies roughly halfway between the plateaus above and the rivers below, a broad bench of hard white sandstone some 250 million years old, the same age and appearance as the sandstones and limestones that line the rim of the Grand Canyon some three hundred miles to the southwest.

By sunset the rain showed no signs of letting up. I was traveling solo and light without stove or tent. I spent the night under a tarp strung around the lee side of a boulder, watching the rain fall in the desert. Flashes of lightning illuminated the ground—periods of total darkness alternating with sudden bursts of bright, clear light—offering brief glimpses of distant canyons and needles of rock. Tired from three days of hard walking, I fell asleep before midnight.

When I awoke the next morning at twilight the sky was dark blue and cloudless. The rain had washed the desert clean, leavings pools of water behind in the pocked and rolling surface of the rock. As the sun rose over the edge of a distant mesa, they sparkled and glittered like jewels, stretching along the canyon rim as far as the eye could see.

Water is life in the desert, not only down in the seeps and springs that dot the canyons below but up here on the barren reach of the rock as well. The flat surface of the sandstone here is scored by cracks and joints—thin lines of fracture that run across the ground in a gridwork of gullies and furrows. They form a natural drainage system for water falling on the bare rock and trap windblown bits of grit and sand. Here and there you can see small, tough plants that have taken root in these narrow seams in the sandstone: solitary clumps of juniper, cliff rose, and aster that seem to draw their life right out of the rock. Elsewhere patches of sand and loose soil are covered by hummocky black crusts. They are not layers of dried mud or silt but cryptobiotic soils—mats of cyanobacteria, algae, fungi, and moss that cover the shallow desert soils with a thin blanket of life, a collection of plants known as cryptogams. Even bare holes and pockets in the rock are a source of life, collecting points for water after the desert's scarce rains. Dry for most of the year, they seem to contain nothing more than loose grains of sand and small bits of debris. When the rains return, however, these swales and depressions are magically transformed into tiny desert seas, inhabited by species of freshwater shrimp and a collection of tiny animals and insects capable of surviving the long dry spells between rains—a swirling cloud of life.

Biologists call these rock pools in the desert potholes or tinajas. They can be found almost anywhere on the Colorado Plateau where flat areas of bare rock are exposed. They range in size from shallow saucers a few feet in diameter to deep bowls large enough to hide a car or small house—smooth-sided tanks twenty and thirty feet deep. In the slickrock deserts of southeastern Utah you can pass hundreds of these potholes in the space of a mile or two. "Potholes are kind of neat from a natural history perspective," said Tim Graham, a biologist with National Biological Survey in Davis, California, who has been studying pothole ecology on the Colorado Plateau for several years. "You add water and, boom, they're up and running. Ecologically they're interesting as well because they are so discrete. There is a very sharp boundary between that aquatic system and the terrestrial system." Although small, the variety of life is bewildering: fairy, ghost, and tadpole shrimp, snails, gnats, tartogrades and rotifers. They feed on algae growing in the water and blown-in bits of detritus—bits of leaves, grass, and lichen—as well as each other.

While some potholes are so small that it is literally possible to count every organism living in them, scientists are only just beginning to understand their ecology. Before moving to Davis, Graham worked for several years for the National Park Service in Moab, Utah. When he first started to study potholes, he assumed that their abundance in the slickrock deserts outside of town would make research easy. With thousands of seemingly identical potholes to choose from, he thought it would be possible to design a whole series of carefully controlled experiments: changing the pH level or acidity of a particular pool, for example, or removing a particular species from it and then comparing the effects to an undisturbed pothole nearby. It didn't take long, however, for him to realize that every pool was different. "Even when they're right next to each other they're different," he said. "For example, there's one that you'll always find tadpoles in and then one right next to it that you'll never find tadpoles in. They look about the same, but it ends up that one of them lasts longer than the other one. I don't know how a toad assesses that when it gets there because the depths are about the same. Maybe one is tucked up under a rock a little more so it doesn't get heated up as quickly. But there are so many questions like that. What is it that makes them do that? What allows them to do that?" As if that weren't complicated enough, Graham said, the composition of each pool varies from day to day as competing populations of predators and prey rise and fall.

* * *

Bare rock is relatively rare on the earth's surface. Most is restricted to extreme environments: rocky coasts, deserts, and high altitude areas above timberline where the extreme cold and high winds make it impossible for trees to grow. Each environment offers its own challenges to growth: the pounding action of waves and the corrosive effects of salt water on a rocky coast, the extreme cold and short growing season in alpine areas above timberline. Here in the desert, life is limited by a chronic lack of water. Life on earth began in the sea and is still inextricably linked to water. It took over a thousand million years of evolution for the world's earliest forms of marine life to adapt themselves to life on land. Some, like the freshwater shrimp that thrive in the potholes of the Colorado Plateau, however, have never really lost their link to the sea. No one knows for sure exactly how old these tiny desert animals are, but fossils of what appear to be tadpole shrimp have been found in rocks that date all the way back to late-Paleozoic time more than 245 million years ago. By the time the first dinosaurs appeared, desert potholes like these may have already been teeming with life.

While other desert animals like lizards and mice still depend on water in one form or another to survive, the shrimp and other tiny pothole animals need not only water to live on, but water to live in as well. That total dependence on water in such a dry and hostile landscape has forced them to adopt a variety of strategies to survive. Snails, for example, simply seal off their shells and sink to the bottom of the pools once they begin to dry, Graham said. While their hard shell keeps water loss to a minimum, the snails themselves enter a kind of dormant state and wait for the rains to return. Shrimp and other crustaceans, however, survive the long dry spells as eggs that are blown and scattered by the wind after the pools dry up. Although small, the eggs are surprisingly drought-tolerant, capable of losing as much as ninety percent of their weight in water while still remaining viable. Even more difficult than surviving the long droughts and intense heat of the desert, however, is deciding when to hatch and come back to life. Rainfall is not only slight here in the deserts of the Colorado Plateau, it is also extremely unreliable. A light spring rain may fill the pothole with only enough water to last a few days. Fall rains can come so late in the year that water in the pools freezes solid in the space of a few days or weeks. For the shrimp to survive from year to year their eggs must have enough time not only to hatch and mature but to lay a new clutch of eggs as well. Timing is everything.

"I like to say these guys put all their eggs in one basket, but they lay many types of eggs," Graham said. There are eggs that hatch as soon as they get wet. Some have to get wet and then dry and wet again, he explained. Others have to get dry and then cold and then wet. Some need a flash of light—a signal perhaps that the sand and grit on the bottom of the pool have been overturned by a new influx of water. They seem to lay eggs for all occasions, continually hedging their bets—ensuring that even after the most promising rain a few eggs will be left behind if the hatch should fail. A single shrimp can lay as many as 500 eggs.

While shrimp must make it from egg to egg to survive, other forms of life found in the potholes such as gnat larvae simply dry up like a sponge when the water disappears, springing back to life when the pothole fills again with water. The larvae are cryptobiotic, capable of surviving in a desiccated state by actually changing their cellular structure as the water disappears.

Water is critical to cells because it holds their proteins and lipids in a three-dimensional structure. As the pools dry up these gnat larvae replace those water molecules with sugar molecules that keep their cells' proteins and lipids in place until water is available again. The process, however, is not without its own demands. It takes time and energy to create these sugar molecules, and if the intervals between wet and dry periods come too close together, the larvae soon run out of energy. And, if they dry out too quickly, they simply get baked. "It's like frying an egg," Graham said. "You can't put it back together again." If all goes well, however, and the larvae make it back to their stable cryptobiotic state, they can survive indefinitely. It's almost as though there was some irreducible core of life locked up inside them—a fact, Graham said, that raises some interesting questions. "How old are you? It's been ten years since you hatched out of the egg, but you've only been wet for three days."

While deserts seem wizened with age, much of their flora and fauna evolved only recently—part of the continuing process of evolution that has seen life progressively evolve to survival further and further away from water. Some 2,500 million years ago plantlike cells of blue-green algae or cyanobacteria were possibly the only form of life on earth. So primitive in design that they lacked even a cell nucleus, they dominated the world's oceans, growing in large mats and mounds on the seafloor known as stromatolites. In spite of their simplicity, scientists believe that they may have reshaped the earth, making it habitable for the varied and sophisticated forms of life we know today. As they carried on photosynthesis, the oxygen released in cell respiration may have helped create the earth's oxygen atmosphere.

Nearly 1,000 million years before the present, these primitive cells of blue-green algae or cyanobacteria seem to have moved onto land as well. Perhaps the daily rise and fall of the tide along the shorelines of ancient seas slowly led to the evolution of newer varieties of cyanobacteria capable of living on dry land. Fossils of terrestrial blue-green algae more than 900 million years old have been found—suggesting that life on land has deep roots as well.

The sudden appearance of hard-shelled marine invertebrates at the start of Paleozoic time some 570 million years ago ended blue-green algae's dominance of the ancient oceans, clearing the way for the evolution of newer, more complex forms of life.

Varieties of blue-green algae still inhabit the earth's surface today, little changed from their first appearance several thousand million years ago. You can find them growing in large mats and mounds on the floors of shallow hypersaline lagoons in shallow desert seas and out on the treeless reach of the Arctic tundra. Pioneering plants, they thrive in areas where few other plants or animals are capable of surviving. Out at ground zero on the Nevada Test Site, scientists have found cyanobacteria growing on the baked dead ground left by the explosion of a thermonuclear bomb. Several million years of evolution have not been able to improve upon their simple design. While it is fascinating that forms of life as elaborate as trees can evolve from something as simple as a single plantlike cell, it is no less intriguing to ponder the fact that some forms of life have managed to remain essentially unchanged for millions or even thousands of millions of years.

You can find varieties of cyanobacteria growing in the deserts of the Colorado Plateau as well, growing in black mats of cryptobiotic soil along with tufts of lichen and moss. Composed predominantly of a plantlike species of bacteria known as *microcoleus vaginatus*, they make up as much as 70 percent of the biomass in places. In fact, not only do they dominate the area, in places these tiny ground-hugging plants or cryptogams seem to be the key to the entire ecosystem—pioneering plants that hold fragile desert soils in place and fix nutrients out of thin air.

One of the first scientists to recognize the role that forms of cyanobacteria such as microcoleus played in the region was Brigham Young University's Kimball Harper. In the 1960s he was flying over what is now the Needles District of Canyonlands National Park when he noticed that the soils were chocolate brown. The color was darkest in remote parklike pockets of ground that had been seldom grazed by cattle or visited by people. He soon went down on the ground to find out what was behind it all only to discover that the deserts were covered with crusts of cryptobiotic soil—mats of algae, moss, and cyanobacteria. For a biologist like Harper it was a surprising find. "If you had to pick a place where blue-green algae would have the greatest impact, it probably wouldn't be the desert," he said. "Life and the stability of sandy surfaces, however, depend on it."

Out in the slickrock deserts of the Colorado Plateau soils are thin and poor. Late-summer thunderstorms send sheets of water pouring across the desert wearing down rocks and washing away the eroded sand and silt they leave behind before they ever had time to be transformed into soil. What little remains is typically sterile or nearly so—grains of quartz worn from rocks that were once barren sand dunes or the salt-laden silts and clays of ancient lakes and seas. Mats of microcoleus help combat this washing away by sending out an almost invisible network of rootlike filaments that hold both themselves and the soils in place. In photographs taken by a scanning electron microscope they seem to wrap around grains of sand and particles of clay like tentacles.

While its filaments reach out through the soil, the microcoleus itself is protected by a sticky sheath. Biologists like Jayne Belnap with the National Park Service in Moab, Utah, believe that its sticky surface may help it hold on to not only grains of sand and particles of clay, but key nutrients as well. While the sheath keeps them from drying out during the desert's long dry spell, the microcoleus spring back to life almost instantly as soon as the rains return. Tiny roots and filaments grow and spread, binding it even more closely to the soil. In places they reach down more than three inches below the surface. Rather than rely solely on the nutrients they can extract from the ground, these tiny threads of cyanobacteria are capable of fixing or extracting key nutrients like nitrogen right out of thin air. In time the stable layers of soil and storehouse of nutrients these crusts of cyanobacteria build up attract plants and other organisms—mosses and lichens that help build up the soil even further.

Although these plantlike cells of bacteria carry on photosynthesis just like grasses and trees, their color is black not green—a trait that seems almost illogical in the desert where the summer heat is almost killing. But while the land here is hot and dry in the late spring and early summer, in the winter and early spring it is often bitterly cold. Microcoleus' black color enables it to heat up rapidly in the early spring and spring into action when most other plants in the high desert are still dormant. On days when the air temperature is barely above freezing, mats of microcoleus can bask in 70-degree heat, Belnap said. Once they begin growing, nitrogen levels in the soil can shoot up by as much as a thousandfold.

But while these tiny forms of cyanobacteria are rugged, capable of withstanding not only winter air temperatures that drop to twenty below zero and summer ground temperatures that climb to more than 150 degrees Fahrenheit, their mats of rootlike filaments are brittle and fragile when dry. A single footprint can undo several hundred years of growth. Biologists are uncertain just how much nitrogen cyanobacteria such as microcoleus can produce, but research has shown that in areas where these cryptobiotic crusts have been damaged or destroyed plants such as sagebrush and cliffrose show nitrogen levels that are 20 to 30 percent lower than those in areas where cryptobiotic soils are healthy. Over the past few decades continued grazing by cattle and sheep and a rapidly increasing flood of visitors have caused these areas of damage to spread dramatically. The signs of overuse here are expressed not just in changing distributions of plants and trees, but also right down at the land's base—the simple cells of cyanobacteria that hold the soil together and gather the nutrients on which all other life depends.

While cryptogams help other plants establish themselves, their presence is not entirely benign. While cyanobacteria like microcoleus help fix nitrogen in the soil, they also compete with other plants for scarce supplies of nutrients like iron and phosphorous. In time desert grasses and shrubs can crowd microcoleus out, making use of the ground it has prepared for their arrival.

In less extreme parts of the world biologists talk of plant succession: the shift from abandoned field to hardwood forest that occurs in the woodlands of the eastern United States or the slow and steady progress from open burn to dense pine forest in the mountains of the western United States. Here on the Colorado Plateau, however, there is no guarantee of an orderly progression from bare rock to pinyon and juniper or desert grassland. In places both the climate and landscape are too extreme for anything but cyanobacteria and their crusts of cryptobiotic soils to grow. Flash floods and rapid erosion are continually resetting the clocks here—taking the land back to zero. Plants here are still waiting for the land to catch up with them. Like the pines and firs hidden away in the alcove forests, these cryptogamic soils are living fossils, an ongoing illustration of the earth's ancient history, settling and resettling the landscape here just as they did nearly a thousand million years ago. Here on the Colorado Plateau the past still mingles with the present.

E I G H T

Hunters and Gatherers

THE ROCK-SHELTER is low, so filled in with sand and silt that its ceiling is little more than three feet off the ground. It sits at the base of Split Mountain a few hundred yards from the paved road that leads past the fossil quarry at Dinosaur National Monument to the campgrounds alongside the Green River. Dozens of pictographs and petroglyphs decorate its walls and the sides of nearby rocks— pictures of deer and antelope and human figures drawn by the Fremont people nearly one thousand years ago. Some are painted on the rocks in shades of ocher and tan. Others have been picked and chipped into the surface of the sandstone with bits of stone. A pre-historic pueblo people like the Anasazi who settled the Four Corners area to the south, the Fremont lived in the northern reaches of the Colorado Plateau, settling in small villages of pithouses and pueblos in the canyons and mesas of Utah and northwestern Colorado. Traces of their villages and rock art are found as far south as Canyonlands National Park in southeastern Utah.

Unlike the Anasazi, who built large pueblos of several hundred rooms at places like Chaco Canyon and Mesa Verde and developed distinctive styles of pottery and architecture that spread throughout the Four Corners region, the Fremont were never so collected or uni-form. Archaeological sites suggest that beyond certain basic traits, styles of architecture, pottery, and even spearpoints varied from region to region and even canyon to canyon. Instead of large pueblos clinging to the sides of cliffs, the Fremont left giant panels of pictographs and petroglyphs scattered across the northern reaches of the Colorado Plateau; stone galleries of pecked and painted rock with figures five and ten feet high, the images of lizards and spirits, even stars and stalks of corn. Sometime between 1300 and 1400 the Fremont people vanished. There is no clear consensus of opinion among archaeologists as to just where the Fremont people went. While some believe they may have moved east onto the Great Plains, others suggest that they headed south to join up with the Anasazi.

Other ancient people appear to have used this rock shelter before them. Less than a mile away the Green River emerges from the center of Split Mountain and begins its meandering journey southward through the flat, open deserts of the Uinta Valley. Standing here at the mouth of the shelter you can look southward across it—a barren and seemingly lifeless desert broken only by the scattered lines of bluffs and mesas. In midsummer the heat and sun are relentless. In winter, storms bring wind-driven clouds of fine dry snow. For the Fremont who decorated these rocks with petroglyphs and pictographs, this small rock-shelter at the base of Split Mountain would have offered protection from the sun and wind, just as it had for those who lived here before them.

Several feet below its present floor archaeologists studying this site have found projectile points, the tips of spears and lances some six thousand to nine thousand years old, the relics of Paleolithic hunters, descendants of the first roving bands of hunters and gatherers to reach North America from Siberia and Asia. If the twisted black schists that lie at the base of the Grand Canyon or the ancient rocks that lie at the core of the Uinta Mountains are part of the continent's stable craton, these spearpoints and knives from Paleolithic hunting camps are the stable base, or beginning, of the continent's human history.

Rock-shelters and caves scattered across the United States and Canada yield artifacts of similar age and design—spearpoints pecked from pieces of flint or obsidian, tools of wood and bone for butchering animals and cleaning hides—suggesting that the hunting and gathering culture of these prehistoric peoples was both widespread and surprisingly uniform. In the rich archaeological record of the Southwest you can see how the mobile lifestyle of these primitive hunters and gatherers slowly gave way to the more settled world of the Fremont and Anasazi as new tools, techniques and crops were slowly discovered and developed. Here in the northern reaches of the Colorado Plateau, however, that hunting and gathering lifestyle of the past would never completely disappear. While the Fremont had begun to abandon this area by the early 1300s, new arrivals would begin to take their place: nomadic peoples like the Ute and Paiute who were living in the mountains, canyons, and mesas here when the first Europeans arrived in the late 1700s. Migrating into the area from the deserts of southern Nevada and southeastern California, they brought with them a mobile hunting and gathering way of life that was almost indistinguishable from that of the Paleolithic hunters who had lived here several thousand years before them.

Man is a fairly recent arrival in North America. While rocks on the floor of the Grand Canyon and in the core of the Uinta Mountains are more than 1,000 million years old, man may have been present in North America for no more than twenty thousand years. While both Europe and Asia are littered with signs of early man more than fifty thousand or a hundred thousand years old, it is hard to find archaeological sites more than fifteen thousand years old in North America.

Less than twenty thousand years ago large portions of New York, Ohio, Indiana, and Illinois were covered with ice, glaciers that had pushed southward into the United States from Arctic Canada. At times the ice reached as far south as Kansas. In places it planed the land flat. Elsewhere it scored the ground with deep gouges and pits. Some were later filled with water to become ponds and lakes. Both the sealike expanse of the Great Lakes that straddle the Canadian border and the maze of lakes and ponds that spread across the northern reaches of Wisconsin and Minnesota were created by glaciers. Advancing sheets of ice more than a mile thick in places moved across the ground like the blade of a plow, pushing mounds of rock and soil in front of them. When the glaciers retreated, they were left behind as terminal moraines marking the furthest reach of the ice. Today they appear as ranges of hummocky hills in places like Ohio and Wisconsin. Farther east along the coast they form peninsulas and islands like Long Island and Cape Cod. The glaciers' reach and power were almost unbelievable. At times so much water was locked up in ice that sea level was more than three hundred feet lower than it is today. In the North Pacific there was no Bering Sea. Siberia and North America were connected by dry land. While this link between continents is commonly referred to as a land bridge, it was not a narrow neck of land but a broad subcontinent several hundred miles wide known as Beringia.

Over the past 2 million years repeated episodes of glaciation have sent sheets of ice moving back and forth across North America, leaving Beringia periodically above water. While fossils and the distribution patterns of current populations suggest that plants and animals have used this alternately appearing and disappearing link between continents several times over the past few hundred thousand years, early peoples seem to have migrated across it only recently, sometime between twenty-three thousand and eight thousand years ago.

While blood type, tooth structure, and genetics among the native peoples of both North and South America show a strong similarity to those of the peoples of northern Asia, they also fall into three distinct groups—suggesting to archaeologists and anthropologists that migration to the New World may have occurred in at least three separate pulses. The first and largest is believed to have occurred roughly eighteen thousand years ago with the arrival of the *Amerind* peoples, who would become the Paleo-Indian peoples of both North and South America. Their descendants today include the vast majority of native peoples in the Americas, including not only most pueblo peoples in the southwestern United States but the Inca and Mayan peoples of Central and South America as well. The second migration brought the *Na-dene* people to northern Alaska and Canada between fourteen thousand and twelve thousand years ago—ancestors of the Athabascan people who lived in the forests of northern Canada. Several thousand years later groups of these Athabascan peoples began to spread westward and to become the coastal Indians of southeast Alaska, British Columbia, and the Pacific Northwest. Others spread southward, reaching the Four Corners area soon after the Anasazi had abandoned it, to become the Navajo and Apache. The third and final migration across Beringia may have taken place as recently as ten thousand years ago with the arrival of the Inuit, or Eskimo, peoples of Arctic Canada and Alaska and the Aleut peoples who inhabited the tundra-covered islands of the Aleutians that stretch across the North Pacific like stepping stones, running from the Alaska Peninsula to the coast of Kamchatka on the edge of Siberia. Both the Aleut and the Inuit are seafaring peoples, and this final migration may have taken place by sea kayak as Beringia disappeared beneath the rising waters of the Pacific.

This migration to the New World was merely the final step of a 3-million-year journey that had brought early peoples out of Africa and onto the surface of every continent except the frozen and almost lifeless world of Antarctica by the close of the Ice Ages. The first upright, bipedal walking man appeared in southern Africa some 2.7 million years ago, Australopithecine, or southern man. Over the next two million years early man would spread throughout Africa. Between 1.5 million and 500,000 years ago, these early peoples would develop the skills needed to survive in the colder lands to the north: discovering fire and the wearing of furs and skins. By 350,000 years ago, early man had begun to move out of Africa and into the Near East, Asia, and China. The total world population, however, was still small and may have numbered no more than ten thousand.

Some two hundred thousand years ago early man had reached Europe. By a hundred thousand years ago they were hunting woolly mammoths and mastodons with spears and lances in Central Europe—the Stone Age or Neanderthal man of textbooks, comic books, and popular imagination. Forty thousand years ago they were pushing out into the plains and steppes of Siberia. By twenty thousand years ago they would reach the edge of the Bering Sea. Their arrival coincided with the last major surge of Pleistocene glaciation that would open up a pathway to the New World.

Quite likely the journey across Beringia to North America seemed no different to early man than the journey across Europe and Siberia that had brought them to the edge of this New World. They may not have even been aware that they were traveling, but simply followed the same herds of mammoths, bison, and antelope they had hunted on the steppes of Siberia. While the ice sheets at times reached as far south as Kansas, they were divided into two lobes:

one that spread southward from the Hudson Bay area to cover the Great Lakes region; another that spread southward from the northern Rocky Mountains and the Coast Ranges of northwestern Canada and Alaska. In between was an ice-free corridor leading south toward the Great Plains and the vast interior of North America. A highly mobile society of hunters and gatherers, these early peoples may have needed only a few thousand years to reach Tierra del Fuego at the southern tip of South America. By eleven thousand years ago, scattered bands of Paleolithic hunters could be found almost everywhere in North America.

Today the campsites and kill sites of Paleolithic hunters have been found not only in the Southwest but in places as diverse as Florida, Wyoming, Pennsylvania, Texas, and California. They fashioned tools from chipped pieces of stone: both delicately flecked spearpoints of flint and obsidian and cruder knives and scrapers for the butchering of game and the cleaning of hides. Others were made out of bone and wood, such as wrench- or dielike devices for straightening the shafts of spears. Not only are such sites widespread, but the artifacts from rock-shelters and caves all over the United States and Canada from this early era yield artifacts of similar age and design, suggesting that the hunting and gathering culture of Paleolithic times was both widespread and surprisingly uniform. That interpretation can be extended worldwide. The tools of Paleolithic hunters in Europe, Asia, and Africa are, in many respects, almost indistinguishable from those of North and South America. Fifteen thousand years ago the world may have been more of a global village than it is today in an era of fax-modems and satellites. Early man was linked not only by blood but by culture as well.

The first Paleolithic hunters to reach the Southwest would have found a world very different from the one we know today. The climate was both cooler and wetter. Glaciers covered highland areas like the Mogollon Rim on the southern edge of the plateau and the Uinta Mountains and high plateaus of Utah to the north. Boreal forests of pine and fir reached as far south as central New Mexico. The Great Plains were a steppelike grassland, a kind of American Serengeti where herds of elephant-like woolly mammoths, long-horned bison, camels, and antelopes were pursued by saber-toothed tigers and dire wolves. Near Folsom, New Mexico, on the high plains that cover the eastern half of the state, archaeologists have found the carcasses of mammoths and mastodons more than fourteen thousand years old with the spearpoints of Paleolithic hunters embedded in their ribs. Points of the same style and design have been found on the northern edge of the Colorado Plateau, relics of Folsom man, as these primitive hunters were known. The animals were apparently ambushed near water holes or driven into swamps and then butchered on-site. A single mastodon carcass could have provided a band of hunters with several hundred pounds of meat.

Neither the climate nor the rich array of animals, however, would last for long. By ten thousand years ago the climate was already beginning to warm as the Ice Ages drew to a close. While boreal forests in the lowland areas of the Southwest were giving way to desert grass and scrub, large animals like mammoths and mastodons were rapidly disappearing. In addition to changing climate, some archaeologists have suggested that early man may have played a role in the extinction of these large Ice Age animals. Unaccustomed to humans, they could well have been easy prey—a perception suggested by the discovery of kill sites containing the carcasses of more than a dozen animals. Such hunting pressures when coupled with the region's changing climate could have proved catastrophic for these large, slow-to-reproduce animals.

With the disappearance of mammoths and mastodons, early man in North America shifted to other sources of food. On the Great Plains bands of mobile hunters began stalking herds of bison. Out on the Colorado Plateau these Paleolithic hunters switched to even smaller game: rabbits, birds, and deer. They not only lived in caves and rock-shelters but also built brush huts and windbreaks. Growing population and a steadily shrinking supply of resources would gradually force them into an increasing dependence on gathered, as well as hunted, supplies of food: nuts, leaves, roots, and berries—setting the stage for the adoption of agriculture and the rise of prehistoric pueblo peoples like the Fremont and Anasazi.

In spite of the resourcefulness and ingenuity of these early pueblo peoples, the high desert world of the Colorado Plateau is a marginal environment for human survival. Less than one thousand years after their appearance in the canyons and deserts of the Four

Corners region, both the Anasazi and the Fremont would be gone. In their place nomadic peoples like the Ute and Paiute would appear, along with the Navajo and Apache. The evolution of people and culture is by no means inevitable or uniform—both of these newly arriving groups would bring with them a way of life that would no doubt have been familiar to the first Paleolithic hunters to reach the New World. In the isolated world of the Colorado Plateau, the past would come back to life.

The history of ancient man in the Old World was intimately tied to biological or physical evolution, to changes in bone structure, physique, and brain size through a succession of different early peoples that progressively led to the form and structure of modern man: *Australopithecus, Homo sapiens, Homo sapiens neanderthalis,* and *Homo sapiens sapiens.* By the time early man reached the New World, however, human evolution was no longer predominantly biological or physical but cultural and intellectual, as new cultures and new ways of life began to evolve.

Biologists often speak of convergent evolution—isolated species of plants and animals growing in similar climates and environments that evolve into similar or even identical forms. What holds true for biology, however, seems barely tenable when speaking of human culture and society. When the first Europeans reached the New World in the late 1400s, they found not only a land they had scarcely imagined but peoples and civilizations unlike any they had ever seen. They were separated from the New World not only by several thousand miles of open ocean but several thousand years of time as well. Out of the simple common base of Paleolithic hunting and gathering societies, several hundred different cultures had emerged. Tied to the varied demands of coasts, plains, forests, and deserts, the native peoples of the Americas had their own distinctive languages, customs, and habits. Some, like the Inca, Maya, and Aztec, seemed almost dreamlike: exotic reflections of the cities and civilizations they had left behind in Europe. Others seemed so primitive that they wondered if they were even human. Some, it was rumored, slept underwater. Others ate no food, but simply lived off its smell. No story was too fantastic to be believed.

Europe, Africa, and Asia, of course, had given rise to a rich variety of civilizations and peoples as well: the ancient empires of Egypt, China, Greece, and Rome. More than three thousand years of conquest and empire building, however, had left few stones unturned in the Old World. What struck early Europeans was not just the differences between peoples and cultures in the Old World and the New but the profound differences between peoples and cultures within that New World as well. While the Inca and Maya had built cities and temples larger and more sophisticated than anything found in Europe at the time, a few miles outside their elegant city walls one could also find people whose way of life seemed almost identical to that of prehistoric peoples who had left their spearpoints, campfire rings, and cave paintings scattered across Europe.

That contrast was found not only in Central and South America but in North America as well. In the American Southwest, bands of nomadic hunters and gatherers like the Ute and the Navajo roamed just beyond the edge of the carefully tilled fields of the pueblos. Coming in contact with these nomadic peoples was, for the Spaniards, like staring across twenty thousand years of time. While the Navajo would eventually learn the cultivation of corn, beans, and squash from the Pueblos and acquire herds of sheep and goats from the Spanish to develop a pastoral culture all their own, groups like the Ute would never take to settlement or agriculture but continue to live as nomadic hunters and gatherers until they were forced onto reservations in the late 1800s.

Time and the evolution of society and culture do not necessarily lead to increasing sophistication and structure. Some ways of living are so tightly tied to the landscape around them that they persist almost unchanged for thousands of years. If the pueblo peoples of Arizona and New Mexico are a modern-day reflection of prehistoric pueblo peoples like the Anasazi and Fremont, nomadic peoples like the Ute are a modern-day reflection of the first roving bands of prehistoric hunters and gatherers to reach the New World.

The Ute have many names. The Zunis called them Deer Hunting Men. The Ponca and Omaha of the Great Plains called them Rabbit Skin People. Others called them the Blue Sky People, a reference, perhaps, to their homeland in the high, clear air of the Rocky Mountains and the northern reaches of the Colorado Plateau. The Ute, however, referred to themselves simply as the *Nuche*, "the people." Those outside the tribe were known only as strangers or enemies. In Spanish *Nuche* would be translated as *Yuta* and from that into English as both Ute and Utah—a name they would eventually lend to the state that now covers much of what was once their homeland. By the time the first Europeans arrived, Ute territory stretched all the way from the front range of the Rockies in Colorado to the Wasatch Front in central Utah, and from the Wyoming border to the frontiers of what are now New Mexico and Arizona.

Unlike Pueblo and Navajo peoples, the Ute have no elaborate stories of their own beginnings; no stories of emergence or migration from previous worlds now hidden beneath the earth's surface. They believe merely that they have been present since the earth's beginning. According to some traditional stories the Ute and other peoples were created by *Sinawaf*, the creator, from a bag of cut sticks. Although the Ute were a small tribe, it was said that the creator decided that they would be the hardest to kill and capable of conquering all the rest.

Archaeologists and anthropologists, for their part, have no clearcut opinion on the origins of their Ute either. Studies of their language and culture suggest that they were part of an archaic desert culture that began migrating out of southern Nevada and southeastern California around 1100 A.D. From the Death Valley area and the deserts of southern California and Nevada, these early hunters and gatherers began spreading outward and northward in a fanshaped pattern. By the time they reached the Four Corners area in the 1400s, the pueblos and cliff dwellings of the Anasazi had already been abandoned for more than a century.

Although the Ute would later become known for their fine horses and tepees, those who first came to the Colorado Plateau and the Four Corners region traveled on foot, carrying what they needed in packs or bundles. At temporary camps they used piles of brush to make shelters and windbreaks. Elsewhere they built domed huts or wickiups out of willow branches, juniper bark, and grass. They lived constantly on the move and seldom stayed anywhere for more than a month, except in the winter, when they covered their wickiups with dirt and skins for added protection from the wind and the cold and stayed put, waiting for the weather to break. Like the

Fremont they created fabulous pieces of rock art—multicolored images of buffalo and Spanish conquistadors on horseback. But beyond that the Ute lived so simply for much of their history that their early camps and settlements are almost indistinguishable from those of the prehistoric hunting and gathering peoples who preceded them. Archaeologically they are almost invisible. Paradoxically enough, most of what we know about early Ute culture comes from the journals and records of Spanish priests and early explorers like John Wesley Powell, who overwintered with the White River Ute band on the Yampa River in northwestern Colorado before starting on his historic trip down the Green and Colorado rivers.

From the beginning, contact with the Ute was sporadic and unpredictable. They had a way of vanishing into the land. Life was a constant search for food and shelter. They hunted mule deer and antelope, capable, it was said, of running them down on foot. Stalking their prey across open terrain, they would curl up like a rock and lay on the ground or crouch motionless like a bush. In winter they would drive a herd of elk into deep snow and then wade up to the exhausted and terrified animals on homemade snowshoes for the kill. Rabbits were hunted in groups, rounded up in corrals and then clubbed to death. No potential source of food was overlooked: snakes, lizards, caterpillars, even swarms of crickets were eaten.

The Ute were as adept at gathering things from the land as they were at hunting. They collected not only fruits and nuts but leaves and roots as well. Most were harvested as they ripened: serviceberry, chokecherry, currants, and rose hips. Seeds and nuts were eaten ripe but also ground into pemmican or toasted with hot charcoal and ash on woven trays. The mixture was kept in constant motion to keep both the nuts and the trays from burning and then winnowed by tossing it repeatedly up in the air, where the wind blew the chaff and ash away, leaving the toasted nuts and seeds behind.

Necessities and luxuries were fashioned from what they could find around them. Pottery broke too easily and was hard to carry. Instead the Ute wove beautiful baskets out of the flexible branches of willows and squawbush. They made hairbrushes from the charred tails of porcupines and used cactus thorns dipped in ash to make tattoos. Although they made arrowheads and spearpoints of their own, they also made free use of those that had been left by those who had lived in the area before them. According to legend the ancient arrowheads and projectile points that littered the ground were the scales of armor from a giant who had been slain at some indefinite time in the past.

In spite of their resourcefulness and ingenuity, like other hunting and gathering peoples the Ute often lived on the edge of starvation. Winters were deadly. When food ran short, they often lived for days or weeks at a time on parched and boiled pieces of leather and buckskin. Elders were treated with respect, but women who lived beyond their childbearing years were believed to turn into witches or snakes and often starved themselves to death. There was no margin for error or luxury.

The acquisition of Spanish horses in the late 1600s changed Ute culture dramatically, expanding both their reach and their wealth. With the help of their "magic dogs," as the Utes first called the horses, their nomadic lifestyle suddenly burst into flower. Ute hunters were suddenly venturing out onto the Great Plains to hunt buffalo among the Comanche and the Arapahoe and pushing northward into Wyoming as well. Instead of scattered groups traveling on foot, the Ute began gathering together in large bands with hundreds of horses. With horses to carry their belongings, temporary wickiups were replaced by finely crafted tepees. They no longer had to leave things behind. Simple skins and furs were replaced by garments made from finely tanned buckskin and decorated with elaborate beadwork.

Their rise in wealth also brought an accompanying rise in power. While the Ute probably numbered no more than eight thousand at their peak in the early 1800s, they became one of the most feared tribes in the Intermountain West, raiding native pueblos and Spanish ranches for livestock and loot. While the Ute roamed far and wide, the more settled world of villages and farms around them became a resource to be harvested like pinyon nuts or deer. Like the Apache and Navajo, they also became active in the Southwest's thriving slave trade—raiding their poorer Paiute cousins to the west for slaves to be sold and traded to other tribes as well as the Spanish in Santa Fe and Taos. A brisk trade in slaves and captives had existed among the warring factions and tribes of the West long before the Spanish arrived, but the lure of Spanish cloth, iron, and wealth would expand it dramatically. In the 1800s wagon trains passing through the area became a source of income as well. For a time they ran the Spanish Trail like a toll road, collecting tributes and fees from supply trains and parties of pioneers. In contrast to the violent imagery of story-books and film, the Ute had far more of an eye for money than murder. They did things with a certain style. In recalling his travels through the West in the early 1840s, the explorer John C. Frémont wrote that the Ute "were robbers of a higher order than those of the desert. They conducted their depredations with form, and under the color of trade, and toll for passing through their country. Instead of attacking and killing, they affect to purchase—taking the horses they like and giving something nominal in return."

The rise of Ute culture in the early 1800s, however, would be quickly followed by its fall. At the close of the Mexican-American War in 1848, lands in the Soutwest claimed by Mexico passed to the United States—and with that symbolic control of much of the land the Ute had come to regard as theirs alone. Although they did not realize it, the world around them was changing rapidly. Now, instead of confronting a scattered and poorly organized collection of Spanish and Mexican settlements, they were suddenly in the midst of an ambitious and rapidly growing country intent on expanding its reach all the way to the Pacific. By the late 1840s Mormon pioneers would be encroaching on Ute land to the west as they spread out-ward from Salt Lake City along the Wasatch Front. By the late 1850s miners and prospectors would be probing the Rocky Mountains to the east, spreading outward from early finds at Cherry Creek and Pikes Peak in search of silver and gold.

Conflict was inevitable, but it was a battle the Ute could only lose. No matter how one might feel about it—angry, apologetic, or merely indifferent—there was no way a tribe of fewer than ten thousand people could hope to hold on to more than a hundred thou-sand square miles of mountain and desert when confronted with a land-hungry nation of several million.

Conflicts and wars between tribes had been lost and won long before the arrival of the first Europeans. The Ute, for their part, had once united with the Comanche to drive the Apache out of northern New Mexico—and the Apache were certainly not the only natives to be driven off their land by other tribes around them. While we tend to think of native America as fixed and stable at the moment of European contact, migrations and shifting balances of power were constant, due not only to changing supplies of game and food but shifting alliances and conflicts between tribes. In the past, however, when one had been driven off a favored piece of land, there was always somewhere else to go, some other sparsely settled reach of ground or some other with whom one could find shelter. By the mid-1800s, however, strangers and enemies were everywhere. In spite of the flowery language and solemn promises of treaties, the land, as always, would belong to those with the power to seize it.

Troubles began first in Utah in 1865 with the start of the Black Hawk War, a series of bloody skirmishes between the Ute and Mormon pioneers that lasted for the better part of two years. By the war's conclusion the Ute would find themselves gathered on an empty reservation in the midst of the Uinta Valley.

In Colorado and New Mexico other bands of the tribe were being pushed as well. The Kit Carson Treaty of 1868 created a reservation of 1.5 million acres in southwestern Colorado and set up a system of Indian agents for the distribution of blankets and food. Southern Ute leader Chief Ouray signed the treaty with the understanding that the "government should strike out all that relates to mill, machinery, farming, schools and going onto a reservation" from the document.

While the reservation was a guarantee that the Ute would have land for their "undisturbed use and occupation," they were soon forced to cede most of their claims to the rapidly growing state of Colorado. When treaty goods failed to arrive, the Ute simply reverted to hunting and gathering as they had done before. Confrontations between settlers and the Ute became more frequent, an accelerating cycle of shootings, beatings, and murders. In Utah the incompetence and corruption of Indian agents had fueled constant conflicts with the Ute. In Colorado those same problems would fuel not just conflicts but an explosion.

Events came to a head in 1878 with the appointment of Nathan C. Meeker as Indian agent to the White River Ute Agency in northeastern Colorado. A reporter, poet, and friend of Horace Greeley,

Meeker had visited the West before and editorialized about the "mental inferiority" of Indians. In his sixties and heavily in debt, he was determined to teach the Ute the values of both "the Christian work ethic" and farming. Insistence that they plow up a favorite horse pasture to plant vegetables soon led to a confrontation. When an argument between Meeker and a local tribal leader escalated into a shoving match, Meeker panicked and called in the U.S. Cavalry. The Ute, many of whom had served as guides for the cavalry in earlier Indian wars, now panicked as well at Meeker's show of force. When the soldiers neared the agency, Ute warriors ambushed them in a narrow valley and kept them pinned down for days. Meeker and the other male employees of the agency were murdered, and the women and children were taken as hostages as the Ute fled into the forests of Grand Mesa. Meeker himself had been clubbed to death in his home and then dragged outside, where a wooden stake had been driven through his mouth, pinning him to the ground.

The attack became known as the Meeker Massacre and kept headline writers busy for days. Neither Chief Ouray nor anyone in his southern Ute band had been involved in the attack, but his skillful negotiation would eventually save the uprising's leaders from hanging. As for the rest, the Ute's fate was sealed. The next year the remaining Ute bands in northern Colorado were ordered to gather near Grand Junction by the following spring. In 1880, under heavy military escort, they were taken to the Ute reservation in the Uinta Valley. Two years before, the few remaining bands of Ute living in the New Mexico Territory near Cimarron and Abiqui had been ordered onto reservations in southern Colorado. Once there, those reservations would steadily shrink in size through a series of legal and political manipulations. "Agreements the Indian makes with the government are like the agreements a buffalo makes with the hunter after it has been pierced by many arrows," Ouray remarked after a treaty conference. "All it can do is lie down and give in." When the last Ute crossed the Green River into Utah in 1880, more than ten thousand years of hunting and gathering had come to a close on the Colorado Plateau.

BOOK THREE

South

MAN'S FUTURE IS even more obscure than his beginnings. To venture
to sound either depth is to enter an unknown, perhaps unknowable,
realm, but it is characteristic of man that he constantly attempts
these journeys.

LOREN EISLEY
1967

NINE

The Green Table

OUTSIDE OF MONTICELLO, Utah, Highway 666 heads south through a rolling desert of grass and sage, scattered with thin stands of pinyon and juniper. Here and there the dry reach of the land is broken by the clean, geometric shapes of cultivated fields: squares and rectangles of alfalfa, beans, and corn, a fertile oasis of farmland sandwiched between the slickrock deserts of southeastern Utah and the San Juan Mountains of southwestern Colorado. Eight hundred years ago the land here was filled with the villages and fields of the Anasazi.

Farther south as the highway approaches Cortez, Colorado, the high cliffs of Mesa Verde rise up out of the desert in a solid wall. Its top is not dry and bare, but blanketed by evergreen forests of pinyon and juniper and stands of ponderosa pine. On top of the mesa, at high points of ground such as Park Point, the view reaches for miles in every direction—out over Utah, Arizona, New Mexico and Colorado, the heart of the Four Corners region itself. To the east and west the mesa seems all but surrounded by mountains—the San Juan and La Plata mountains to the east and the solitary rise of Sleeping Ute Mountain just off the mesa's western flank. To the north and south, however, the desert seems almost endless, a barren landscape of scrub and dry grass stretching as far as the eye can see.

In late summer you can stand near the mesa's crest and watch thunderstorms drifting across the deserts of the San Juan Basin in New Mexico to the south, trailing sheets of rain and fingers of lightning. Gusts of wind leading the clouds sweep the ground before them, stirring up smokelike clouds of dust and sand as they speed across the desert.

Mesa Verde is Spanish for Green Table. Although its top is green, the name is somewhat deceptive for Mesa Verde is not really a mesa at all, but a cuesta, a sloping tableland of rock that tilts gently to the south. Looking down from the high rim of the mesa you can see how it rises up out of the vast deserts to the south like a gently sloping ramp. Canyons reach back into the tilting slope of the land like probing fingers—pathways that lead out of the dry land below and up to the green reach of the land above.

For nearly one thousand years the canyons and mesa top here were home to the Anasazi. They abandoned the area nearly seven hundred years ago, vanishing in the desert only to reappear farther east along the Rio Grande and to the south in the pueblos of Hopi, Zuni, and Acoma. Their settled way of life did not appear here out of thin air, but slowly developed as their ancient ancestors took to planting gardens and tending fields. Two thousand years ago primitive hunting and gathering people were living here around the flanks of Mesa Verde just as they were farther north in the Uinta Valley. The transition from hunting and gathering to farming, however, was slow. The first signs of Anasazi life did not appear at Mesa Verde until some 400 A.D. as bands of prehistoric peoples began moving into the canyon bottoms and up onto the top of the mesa to farm. Like farmers in the deserts below today, they raised crops of corn, beans and squash. They lived alongside their fields in simple pithouses and brush huts, weaving beautiful baskets and trays. Later these would give way to more elaborate designs: finely crafted pieces of black and white pottery and carefully designed pueblos of stone and adobe.

The appearance of these ancient farmers here was not due to accident or chance. While the mesa's half-mile rise waters its crest with several more inches of rain per year than falls in the deserts below, its south-facing slopes also capture the sun's warmth, giving it a growing season that is several weeks longer than neighboring

Montezuma Valley, which surrounds the small town of Cortez just ten miles away and some two thousand feet below. At night the cold air on top of the mesa sinks into the canyons and flows out into the deserts below like water pouring off the slope of a steeply pitched roof.

By 1100 A.D. the Anasazi were building pueblos of several hundred rooms at places like Chaco Canyon and Aztec in the broad desert valleys and dry washes of San Juan Basin to the south. Here at Mesa Verde, however, the Anasazi did not keep their villages and towns near their fields on top of the mesas or down on the canyon floors, but moved up into the walls of the canyons themselves— building pueblos of 50 and 100 rooms in sides of the cliffs, accessible only by ladders and precarious trails with handholds and footholds cut right into the bare rock. It lasted no more than two centuries. By 1290 A.D. Mesa Verde and its Cliff Dwellings were abandoned—the mesa and its canyons as solitary and vacant as they had been more than ten thousand years before. Protected by the overhanging cliffs of sandstone and preserved by the dry desert air, the ruins of the civilization the Anasazi left behind would last for centuries.

Europeans would travel around the periphery of Mesa Verde for nearly two hundred years before its hidden cliff-dwellings were discovered by outsiders. In 1776 two Spanish Friars, Anastasio Dominguez and Silvestre Veléz de Escalante, led a small company of explorers past Mesa Verde as they headed north from Santa Fe looking for a route to California. Although they noted the presence of Indian ruins in the area in their journals and reports, they made no side trips up into Mesa Verde itself. Nearly 100 years later members of the Hayden Survey, one of several far-reaching mapping and scientific surveys launched by the federal government in the West in the years that followed the Civil War, came to the Mesa Verde area to investigate rumors spread by prospectors working in the area of "the marvelous cities of the cliffs." They did not explore Mesa Verde, but focused on canyons and washes in the deserts below where the miners and others had reported finding ruins. They found several clusters of ruins in McElmo Canyon just west of Cortez and spent time studying and photographing them. Two of the survey's members, geologist William Holmes (for whom Mount Holmes in the

Henry Mountains would later be named) and photographer Henry Jackson (whose photographs of the Yellowstone area would prompt Congress to establish it as the nation's first national park), spent months building detailed scale models with clay and plaster casts of the ruins they discovered for the 1876 Centennial Exposition in Philadelphia. Their exhibit on the Southwest's "Aztec Ruins" won a bronze medal and attracted almost as many visitors as Alexander Graham Bell's telephone.

Five years later two cowboys from Mancos, Colorado, Bill and Al Wetherill, discovered Cliff Palace, the largest cliff dwelling at Mesa Verde, while looking for lost cattle during a snow storm. They were working their way along the edge of a canyon when the clouds suddenly cleared, revealing in the cliffs below what looked almost like a lost city. Building a makeshift ladder to reach the ruins, they found rooms full of pots and tools and scattered pieces of clothing—as if their occupants had simply walked off and left. The next day they came back and began exploring other canyons on top of the mesa. In less than two weeks they had found most all of the mesa's major ruins. Their initial discovery was not entirely accidental. The Wetherills had a keen interest in ruins and artifacts. Running herds of cattle from their ranch in southwestern Colorado, they had regularly stumbled across the ruins of small pueblos and panels of rock art. Finding them soon became something of a hobby. When they had a few hours to spare they began riding around looking for ruins. They were on good terms with the Utes, who controlled Mesa Verde at the time, and had been given permission to run their cattle in the area during the winter months. Friends in the tribe had also told them that there were giant ruins up on top of the mesa—although the Ute themselves seldom visited them. For the Wetherills the attraction was almost irresistible.

The Wetherills were soon making as much money carrying pack trains of tourists up to the cliff dwellings at Mesa Verde as they were running cattle. (In fact their ranch would ultimately go bankrupt, a fairly common occurrence in the Intermountain West, where the land was poor and the debts were high.) It took them nearly two decades of pleading and lobbying with state officials in Denver and federal officials in Washington to gain protection for the area. In the meantime the Wetherills would play a leading role in discovery and excavation of dozens of major ruins in the Four Corners area. In 1920 Mesa Verde became the first national park in the United States dedicated solely to the preservation of cultural resources. The cliff dwellings at Mesa Verde became a source of national pride—a sign that the United States had a culture and history of its own apart from Europe. Spread beneath curving roofs of sandstone, the abandoned stone villages of the Anasazi had an almost magical appeal— even across the barrier of more than six hundred years of time.

After its discovery in 1880, Mesa Verde would become the center of an exploding interest in the culture of prehistoric peoples in the United States. Presidents and celebrities, Teddy Roosevelt and Charles Lindbergh among them, came to the Four Corners area to watch archaeologists and excavators at work and see the ruins of pueblos and kivas for themselves. Just as the region's colorful mesas and buttes would lead to a distinctly American school of geology, so too would the region's well-preserved cliff dwellings and pueblo ruins give rise to a distinctly American school of archaeology as workers and researchers uncovered the remains of dozens of prehistoric cities and towns, not only at Mesa Verde, but in other parts of the Four Corners area as well: Chaco Canyon, Canyon de Chelly, and Hovenweep. Early workers such as Frank Cushing, Jesse Fewkes, and Adolf Bandelier excavated literally hundreds of sites. Although most were self-trained, they kept detailed records and accumulated valuable stores of artifacts, laying the groundwork for those who would follow. Most related their findings to the present-day pueblo cultures they saw around them. To the outside world their work offered a glimpse of the sophistication and antiquity of Native American society. Taking place in an era when tribes such as the Ute, Navajo, and Apache had only recently been driven off onto reservations, their work helped change the country's perception of its native peoples—bringing not only the beauties of their past to light but their struggles with the present as well.

South of Mesa Verde, the high, dry world of the Colorado Plateau stretches for more than 200 miles, all the way to the edge of the low, hot deserts of southern New Mexico and Arizona. While the northern edge of the Colorado Plateau is defined by the Uinta Mountains, its southern edge is defined by the Mogollon Rim that runs across central New Mexico and Arizona. A tangled highland of mesas and mountains capped by flows of lava and volcanic rocks, its edge has been defined by a combination of faulting and erosion. Like the Uintas and the Uncompahgre Plateau to the north, the Mogollon Rim area has been cut repeatedly by episodes of mountain building and uplift.

In between this rim and the land to the north is a maze of flat-topped plateaus, mesas, and buttes: the Kaiparowits and Rainbow plateaus in Utah, the Zuni Plateau and Chaco Mesa in New Mexico, the Rainbow, Kaibab, and Kaibito plateaus in Arizona, along with several dozen others. Some were formed by erosion when layers of rock were stripped away to leave solitary pillars and fins of rock behind. Others are underlaid by deep faults and buried folds that forced their way through the Colorado Plateau while the modern Rockies were taking shape farther east.

Beneath the typically green and forested reach of these plateaus and uplifts are flat valleys and deserts—vast seas of grass and scrub like the Painted Desert of northeastern Arizona and the San Juan Basin of northwestern New Mexico. While they call to mind the cool grasslands and steppes where Paleolithic hunters once stalked herds of mammoth and bison, most of the plants and animals here today originated farther south and migrated into the area after the close of the Ice Ages when the climate began to warm. Unlike the forests above, the plants here compete not for space and sunlight but water.

Like the swirling clouds of life found in a pothole or a mat of cryptobiotic soil, the plants and animals of these grasslands and scrublands have learned to adapt to the desert's extremes of both hot and cold. While the landscape here seems bare, it conceals a wealth of detail and artful adaptation. Roots and seeds lie hidden beneath the surface waiting for the first sign of rain. Clay and salt-laden soils have given rise to unique species of plants found nowhere else on earth. Although these grasslands and scrublands cover perhaps a greater area of the plateau country than any other habitat, they are seldom visited or studied in detail; a landscape to be passed over; a backdrop for the more colorful world of deep canyons and high plateaus.

From simple beginnings in dry desert valleys like these, pueblo people like the Anasazi would arise. While man first arrived here from the north, the plants that would eventually reshape both their lives and the world around them had their beginnings farther south—varieties of corn, beans, and squash that were discovered and developed in Mexico several thousand years ago. Although the land here is dry, it is also high and cold. Agriculture hinges on a delicate balance between rainfall and warmth. For the prehistoric pueblo peoples, life was a continual search for the proper time and place.

T E N

Faults and Folds

WE LEFT MOAB just after sunrise, flying southwest across the
Maze district of Canyonlands National Park, an intricately
carved labyrinth of red, white, and chocolate-colored rock, its
varied layers arranged like the strips of color in a serigraph painting.
It was early April, the second day of my flights around the Colorado
Plateau with Bill Dickinson. Leaving the canyons behind, we headed
due west for the Henry Mountains, speeding across the Burr Desert
and the Blue Benches, a barren landscape of low clay hills. Washes
and drainages spread out across the desert below like the veins in a
leaf, winding and meandering through a world as gray and empty as
the surface of the moon. Almost nothing grows on the unstable surface of the clay. Clumps of saltbush and grass are so widely spaced
that you can pick out solitary plants from one thousand feet above the
ground. Here and there the rippled and rolling surface of the ground
below is dotted with the conical mounds of giant anthills—the only
sign of life.

The rocks below are all Mancos Shale, the remains of an ancient
sea that reached all the way from the Arctic Ocean to the Gulf of
Mexico and covered much of the western United States and Canada
with water some 80 to 90 million years ago. At the time the Rocky
Mountains were nowhere to be found. The Ancestral Rockies had

been worn almost completely away, and the waters of this ancient sea reached eastward for several hundred miles. Today layers of Mancos Shale are found as far away as the Great Plains in Kansas and Nebraska. Here on the Colorado Plateau they reach all the way from the high plateaus of central Utah to the edge of the Rio Grande, scattered over the surfaces of deserts and mesas, a blanket for the more brightly colored rocks below.

Heading west across the desert the tilted rocks of the San Rafael Swell seem to rise up out of the ground like a giant wall. Near the Henry Mountains it seems to disappear, curving off to the west and then diving back down into the earth. As we bank in a tight turn around the northern edge of the mountains, however, another giant bend of rock seems to rise up out of the ground to take its place— the Waterpocket Fold. It runs due south toward Lake Powell between the Henry Mountains and the Aquarius Plateau through a wilderness of slickrock and barren clay. Smooth layers of Navajo Sandstone rise up out of the ground like the crest of a giant wave— a wall of white rock running through the desert for more than fifty miles—its top more than a thousand feet above the dry land below. Streams and washes cut right through the center of the fold in winding slot canyons that expose the brightly colored rocks below. On top are giant holes and tanks worn right into the sandstone— potholes twenty- and thirty-feet deep that hold pockets of water for weeks after the spring and late-summer rains. Down in the canyons are groves of spruce and fir—forests of giant trees left over from the Ice Ages. The hidden forests, Wells tells me, are home to rare Mexican spotted owls. Most are all but inaccessible, visible only from the air above.

Like the San Rafael Swell to the north, the Waterpocket Fold is a broad domelike bend of rock. Its center has been almost completely eroded away. In the flanks of the fold the same rocks that appear in horizontal layers in the cliffs and mesas of Canyonlands National Park to the east are tilted almost on end: vertical layers of Entrada, Carmel, Navajo, Kayenta, and Wingate sandstones—as if a slice of rock from a canyon wall had been cut off and laid on its side.

The land here has not been crumpled up like an accordion but pushed up from below. At the core of the Waterpocket Fold is a deep, almost vertical fault. It does not break through to the surface.

Instead it has caused the rocks above it to bend or bow, like a pencil left between the pages of a closed book. Similar faults and folds are scattered across the surface of the Colorado Plateau. They underlie not only the San Rafael Swell and the Comb Ridge Anticline as mentioned before, but other prominent features of the landscape here as well: the Zuni Mountains in New Mexico, the Defiance Plateau in Arizona, even the Kaibab Plateau where the Colorado River winds through the Grand Canyon. They run roughly north—south across the surface of the Colorado Plateau—almost like sea swell.

These faults and folds are all part of the Laramide orogeny that built up the Rocky Mountains some 50 to 70 million years ago. Four hundred miles to the east faults of the same style and trend slice through the Rockies, but while faulting and folding there built up high jagged peaks, here it merely laced the ground with a series of broad folds and uplifts like this one.

In the decades that followed John Wesley Powell's trip through the canyon country of the Green and Colorado rivers in 1869, the Colorado Plateau became a center of geologic research in the United States. Working in unknown terrain with unbelievably detailed exposures of rock, they developed conventions of geologic mapping still in use today, as well as theories of mountain building, faulting, folding and erosion that literally turned the field of geology on its head. Today, however, the Colorado Plateau is valued more as a classroom than a laboratory for new discoveries. While the brightly colored rocks of the Colorado Plateau offer a wealth of geologic detail, they have little or nothing to say about the movements of the earth's crust that have created its continents and ocean basins. For most of its more than 2,000-million-year history, the Colorado Plateau has been quiet and calm—too far east to be affected by the disturbances that shaped the Basin and Range and too far west to be affected by those that built the Rockies. While the Colorado Plateau and its intricately carved layers of rock gave rise to a distinctively American school of geology in the late 1800s, it was left behind by later theories of plate tectonics that saw the earth's surface as mobile and constantly changing.

While the broad details of plate tectonics were being worked out with regard to the earth's continents and coastlines in the 1960s and 1970s, in fact, skeptics often cited the flat-lying rocks of the Colorado Plateau region and the high rise of the neighboring Rockies as evidence against it, claiming that the new theory, at least with regard to the vast interior of the continent, had more to do with the fertile imagination of its proponents than cold, hard facts. Plate tectonics, they argued, was nothing new. What were seen as plate movements were nothing more than the reactivation of ancient fault zones—lines of fractures and faults that they called linea-ments. Most, they said, dated all the way back to the Paleozoic. The Rockies in Colorado were a case in point. The modern mountains were simply a re-creation of the ancient mountain range that had risen up at the same place several hundred million years before. Drifting continents and disappearing ocean basins played no role in their history at all. Look at the earth's major geologic features, they claimed, and except for a few details, you could relate them all to these so-called Paleozoic lineaments.

That interpretation of things works quite well in the Rockies of Colorado, Dickinson admitted. Study the patterns of faults and folds left by the Ancestral Rockies and the modern mountains today, "and you think you're off to the races," he said. Follow the mountains up into Wyoming, however, and that theory quickly falls apart. The similar Paleozoic lineaments run through Wyoming, but the Rockies there ignore them completely. Utah presents its own problems as well, "small details" like the Uinta Mountains that rise up at right angles to the mountains around them. Instead of following the pat-tern of Paleozoic lineaments, they follow even older Precambrian ones. While ancient patterns of faults and folds are often used by later geologic events, their reuse is not inevitable, Dickinson said, but dependent upon the right orientation and angle of forces. While ancient faults and folds have played a role in defining both the bor-ders and the surface of the Colorado Plateau, so too has the earth's shifting continents and ocean basins. Understanding how plate movements several hundred miles away have shaped the Colorado Plateau and the neighboring Rockies, however, is not a simple problem.

The earth is almost perfectly spherical. In spite of its mountains, canyons, and coastlines, if you could reduce it down in size to the dimensions of a billiard ball, its surface would be as smooth and regular as that of a cue ball. Inside, however, the earth's interior is not uniform but divided into concentric layers: a solid metallic core of nickel and iron surrounded by a fluid mantle of molten rock and all of it capped by a lighter and yet more rigid crust. That rigid outer skin is not solid and stable but broken into fragments, large plates, perhaps more than a dozen in all, that cover the earth's surface like the carapace of a turtle. Floating on top of the more fluid layers below, they have not remained fixed but drifted and shifted through time. Where they pulled apart, they formed ocean basins. Where they collided, they formed mountain ranges or deep ocean trenches.

Before the advent of plate tectonics, it was thought that mountains simply rose and fell. Seas advanced and retreated. There was no unifying theory to tie them all together. In geologic terms the earth's history was disjointed, a collection of effects without any apparent causes.

The rise of plate tectonics changed all that, appearing to tie the earth's varied features together. The rough fit between the coastlines of North and South America and those of Eurasia and Africa were due to more than just coincidence. It could be explained by rifting that had driven the continents apart to create the Atlantic Ocean some 200 million years ago. The continents not only *looked* like they had once been linked together, they had actually once *been* linked together.

While a pulling apart of continents had created the Atlantic, elsewhere their collision had created mountain ranges. The Himalayas, for example, had been created when India collided with Asia. The volcanic peaks of the Andes, in turn, had been created where a colliding plate had not hit the edge of the continent head-on but slid beneath it, thrust down into the earth through a deep trench known as a subduction zone. At depth this descending slab of crust began to melt, creating pools of magma that rose to the surface to fuel volcanoes. Elsewhere along places like the San Andreas Fault, two plates of the earth's crust simply slid past one another, rubbing shoulders as it were, triggering earthquakes.

None of those dramatic features managed to work their way into the Colorado Plateau. "There are no subduction zones here, no magmatic arcs," Dickinson said, "none of the real heat and noise of plate tectonics." Rather than being directly shaped by the plate move-ments taking place around it, the plateau seems to reflect things only indirectly, its surface rippled and creased by more dramatic events taking place hundreds of miles away. Understanding how the plateau fits into the concept of plate tectonics and why it has remained so stable in spite of the intense deformation around it is a tricky problem. "People still don't know how to explain it. They're still arguing about it," Dickinson said. "I think it's actually less com-plex than any area around it. Perhaps that's the problem. It's more difficult to understand why something is less complicated. The threads of details here are easier to see, but it is harder to gather them together." Instead of direct collision between plates, the forces that shaped the landscape here are often indirect and second-hand, part of what Dickinson calls a "subtle intraplate deformation," and much more difficult to understand.

One hundred and fifty million years ago the West Coast of North America was bordered by a subduction zone much like the one found off the coast of South America today. When rifting began to open the Atlantic in the east, the western edge of North America began moving through the Pacific like the blade of a plow, scraping up frag-ments of continents and island arcs to build up California and Nevada. As for the seafloor, however, it did not collide with the edge of the continent head-on, but began to dive beneath it in a subduc-tion zone. Melting at depth, these rocks then rose to the surface as molten plumes of lava and granite to build up the Sierras of California and the Cascades of the Pacific Northwest farther north.

Eighty million years ago, however, the angle of that subducting plate of oceanic crust began to shallow. At first the front of volcanic activity simply shifted eastward, but eventually the angle of subduc-tion became so shallow that the plate of oceanic crust being thrust beneath the edge of the continent no longer penetrated the mantle below. Instead of melting, the rocks began pushing inland beneath the earth's surface like a splinter or wedge, causing the ground above them to buckle and fold. Some seventy million years ago these deep plate movements from the west began to disturb the surface of the Colorado Plateau, bending the rocks into broad domes like the Waterpocket Fold or the San Rafael Swell—the start of the so-called Laramide orogeny. Farther east, however, where the Ancestral Rockies had run, the effects of these deep plate movements were even more dramatic—creating a mountain range that reached all the way from New Mexico to Canada, its tallest peaks nearly three miles high.

Geologists are unsure just why that subducting plate changed its angle of descent during the Laramide. Everything from turbulence within the mantle to changing speeds of plate movement have been suggested. From 80 to 40 million years ago plate motion along the Pacific Coast was some 15 centimeters per year, more than three times as fast as its current speed. Roughly 40 million years ago the plate began to slow down as the Laramide orogeny drew to a close. As the plate slowed and steepened, eruptions of lava and intrusions of molten rock began to reappear, sweeping across the Intermountain West. Like the faulting that preceded them, this sudden resurgence of magmatism would leave the Colorado Plateau largely untouched as well.

While the broad folds and shallow basins of the Laramide were slowly eroded away and covered up, they would never completely disappear. Instead they would remain just beneath the surface, waiting for a new round of disturbances to lift them up toward the sky.

ELEVEN

Grasslands and Scrublands

I N THE EARLY 1970s Kimball Harper was in the midst of a ten-year study of the deserts and badlands that border the Four Corners Power Plant in northwestern New Mexico's San Juan Basin when the desert was unexpectedly favored with a year of record rainfall. Bare rolling hills of sand and clay were suddenly carpeted with thousands of tiny blue flowers. Thickets of saltbush and dry scrub suddenly turned rich and green. Populations of jackrabbit and antelope exploded. Harper, whose study had also included a year of record drought, could hardly believe his eyes. "It was like you were creating antelope out of dust and old rags," he said.

Reaching southward from the foot of the San Juan Mountains in Colorado all the way to the edge of the high, forested plateaus of the Mogollon Rim that mark the southern edge of the Colorado Plateau, northwestern New Mexico's San Juan Basin is a vast world of grass and desert scrub, broken only by gray, bare, clay-colored hills and the scattered lines of low bluffs and cliffs. While the tops of the Colorado Plateau's mesas and plateaus are typically green and forested, the valley floors that lie 6,000 feet below are generally covered by desert grasslands and scrublands, wherever soils are thick enough to give them a foothold. While the high peaks of the San Juans to the north reach up to 14,000 feet, the floor of the San Juan Basin is typically no more than 5,000 or 6,000 feet high. Almost anywhere else in the United States its mile-high reach would transform it into green and verdant highland. Here on the Colorado Plateau, however, it appears as a high, dry reach of desert. Grasslands and scrublands like these are found not only here in northwestern New Mexico, but also in almost every corner of the Colorado Plateau: the Uinta Basin of Utah, the Painted Desert of Arizona, and the Piceance Basin of Colorado.

Heading west on Interstate 40 from Flagstaff toward Albuquerque you can see how the evergreen forests of ponderosa pine and the woodlands of pinyon and juniper that cover the Coconino Plateau slowly give way to grass and scrub as the highway drops down into the Painted Desert. An hour east of Flagstaff you can pull off the highway and walk through the desert, studying the shapes and lines of distant mesas and buttes, fifty, sixty, and even seventy miles away. They seem to float on the edge of the horizon like clouds. The openness of the landscape is startling, a counterpoint to the enclosed world of the canyons and mesas that characterize other reaches of the Colorado Plateau. It gives one the feeling of standing not in the continent's center, but somewhere near its edge, looking out over an endless sea. From the edge of the desert east of Flagstaff, this sea of grass and scrub reaches eastward almost uninterrupted across the southern reaches of the Colorado Plateau through Winslow, Holbrook, Gallup, and Grants all the way to the edge of the Rio Grande more than 400 miles away.

The surface of the Colorado Plateau is broken by hundreds of canyons and plateaus; between them, however, are several hundred square miles of rolling open desert covered by a patchwork of grass and scrub. Although they seem uniform, their details are subtle and shift from place to place with both changing soils and changing patterns of rainfall. In rare undisturbed areas of thick soil there can be stands of almost pure grass: needle-and-thread, blue grama, galleta, and blue bunch wheatgrass. Elsewhere they are intermingled with tough, dry shrubs: thickets of blackbrush, bitterbrush, and sagebrush that mingle with the grasses and sometimes overtake them as well. In places this patchwork of grass and scrub climbs up into the forests above as well, covering the ground beneath stands of pinyon and juniper and the floors of clearings in the forests of ponderosa pine and gambel oak higher up. While the forests of pine and fir that cover the tops of high plateaus and peaks above originated in the boreal forests of Canada, the deserts of scrub and grass that cover the valley floors below originated in the dry mountains and hot deserts of Mexico, migrating into the area after the close of the Ice Ages as the land grew successively warmer and drier. In response to changing climate, they have moved back and forth across the landscape several times. Their most recent advance began some 10,000 to 15,000 years ago after the close of the Ice Ages. A few have more exotic origins: needle-and-thread grass from the well-watered world of the Pacific Northwest and sagebrush whose earliest relatives first appeared in the steppes of Central Asia. Both have been able to successfully adapt to the high desert's extremes of hot and cold.

While deep canyons and alcoves of the Colorado Plateau harbor hidden forests of pine and fir that have survived almost unchanged since the Ice Ages, the grasslands and scrublands here are the products of constant change and evolution. Within them are broad reaches of bare clay hills and salt-laden soils, formed by the erosion of flat-lying layers of Mancos Shale that cover the ground like a blanket. The leavings of an ancient sea, they are so laden in places with salt that they are almost poisonous to plants. The ground in these clay badlands is often unstable as well. After a hard rain the clay-covered ground becomes so slick that in places it seems almost greased. Hills and slopes seem to almost melt in the rain. Drying out simply exchanges one problem for another. There is so much salt in the soil that the ground begins to puff and swell as needles of salt take shape within it. "Walk around out there in May or June and you will sink down right up to your ankles," Harper said of the clay badlands that dot New Mexico's San Juan Basin and Utah's Burr Desert. Harsh as this environment is, there are dozens of plants that have managed to eke out a living here: species of saltbush, aster, and shadscale are found nowhere else in the world. Few perennial plants are capable of surviving here. Instead, most are quick-growing annuals, their long-term survival depending on seeds that wait for the right combination of warmth and moisture to grow.

Elsewhere, even where the soils are deep, the plants and animals of these low valleys and plains carry on their own struggle for survival. Snow carpets the ground here in midwinter, but in summer the desert is hot and dry. Walk across a rolling plain of grass and scrub and you will find cacti scattered across the ground: the thorn-covered shapes of prickly pear and chain-fruit cholla, capable of storing up several weeks' supply of water within their green, waxy skins. Other plants here are no less resourceful in their adaptations to the desert. Grasses here do not grow in carpetlike sod as they do in the greener reaches of the Great Plains farther east, but in solitary bunches separated by patches of bare, open ground. Their spacing is not accidental. Even in a good year, the land here seldom receives more than ten inches of rain. The ground is simply too dry to support the close-packed stands of grass that characterize grasslands in areas such as the Great Plains where the rainfall is higher. They save water in other ways as well. Bladed varieties of grass have rows of flexible hinge-cells running parallel to their central vein that enable them to fold and curl, reducing their surface area to conserve water when the weather turns hot and dry.

While most desert grasses here are seasonal and die off when the rains depart and the temperature begins to rise, the shrubs are often evergreen. Some, mimicking cacti, have narrow, thick-skinned leaves that are coated with wax to reduce water loss. Others have leaves covered with fine hairs, almost like felt, that break up the flow of desiccating winds and may help shade their surfaces as well. Not only do they rely on rainfall from above, but some, like sagebrush, have deep taproots that can reach down twenty feet or more to the water table below. While sagebrush also secretes chemicals that hold plants in check, other plants here like broomrape and Indian paintbrush are parasitic, tapping into the roots of other plants for the food and water they need to survive.

Bare and lifeless during the long dry season, an early spring rain or a late summer thunderstorm can bring these scrub and grass-covered deserts suddenly into flower: purple and blue fields of lupine and foxglove, red patches of Indian paintbrush, orange pools of mallow, and white plumes of yucca and carpets of sego lilies. These bright bursts of color, however, last only a short time. After a week or two the ground begins to dry and the flowers wither and die. Green tufts of grass turn bright gold, set off by the pale green and gray leaves of the blackbrush and sagebrush that surround them.

Life here is hard, not only for plants, but animals as well. Those that live here are either hidden or quick. Black-tailed prairie dogs live beneath the ground in carefully organized towns of perhaps more than one hundred, with an underground network of burrows and tunnels and an aboveground network of rotating watchers who scan the horizon for hawks and coyotes. Others like antelope live entirely aboveground in small herds, capable of running at speeds of 60 to 70 miles per hour. Colored pale tan, their shoulders and necks are marked by patches of black and white hair that serve as a kind of camouflage, breaking up their profile and making them harder for would-be predators to see. They have an unusually keen sense of smell and protuberant eyes that give them an exceptionally wide field of view. Within four days, a newborn kid can outrun a man.

Hair and feathers not only keep these high desert animals warm in winter, but shield them from the sun as well, reducing water loss and insulating them from the heat. Many drink no water at all, deriving what little moisture they need from their food—plants, insects, and even other animals. A kangaroo rat can live indefinitely on a diet of dry seeds, producing the water it needs metabolically in the digestion of the carbohydrates and fats they contain. Heat, however, is as much a problem as water here, and many are active only at night when the desert is cool. Darkness also shields them from predators. Others, like jackrabbits, have large ears that not only give them an acute sense of hearing, but function as radiators as well, laced with a network of tiny blood vessels that enable them to give off heat. Pursued by a coyote, they can run across the ground at speeds of up to 45 miles per hour, covering the ground in 15-foot leaps, spy-hopping or leaping into the air from time to time to scan the terrain ahead.

The grasslands and scrublands of the Colorado Plateau seem almost untouched. Present-day cities and towns cling to the sides of rivers and the well-watered flanks of peaks and high plateaus. Canyons and mesas are dotted with the abandoned villages and camps of pueblo peoples like the Fremont and the Anasazi. Out in the midst of the desert, however, you can walk for hours without seeing a single sign of man. Much of it is open range, devoid not only of ranches and farms but fences and wires as well. Only the Navajo seem to have taken root here, their solitary hogans and corrals scattered across the basins and valleys of their reservation in northern New Mexico and Arizona like islands in a vast desert sea.

In point of fact, however, more than a century of grazing by the herds of both cowboys and Indians has changed the composition and character of the land here almost completely. "Those plants out there are as man-made as plastics," said Kimball Harper. The desert grasslands and scrublands of the Coloradro Plateau are not what they seem.

The first grasses appeared some 70 to 80 million years ago, evolving, paradoxically enough, from trees. Grassland animals appeared soon afterward, and over the next 70 million years they evolved almost side by side, each one shaping the other. Grasses are hard to both chew and digest. Their leaves and stems, in comparison with other plants, are laden with silica that quickly wears away teeth and a high cellulose content that makes them almost indigestible as well. To overcome these obstacles, grazing animals like bison, antelope, and cattle have a number of highly specialized physical adaptations. Their molars, or cheek teeth, for example, are not only highly resistant to abrasion, but successively rise up out of their gums as their upper surfaces are worn away. To aid digestion they have a series of storage chambers or fermentation chambers that precede their stomach where swallowed food can be stored and fermented. To further aid digestion, undigested material is often partially regurgitated as cud and chewed again—a process known as rumination. Where large grazing animals have had a long history on the land—out on the Great Plains or the grasslands of Central Asia or Africa—they have fostered important adaptations of the grasses themselves, bringing about the selection and development of plants that not only tolerate the pressure of grazing but actually thrive on it.

Man has deep ties to the grasslands as well. During the Ice Ages the grasslands of North America, Europe, and Asia were swarming with a host of large animals: mastodons, camels, antelope, and bison. These animals were a source of food for roving bands of Paleolithic hunters. Later, when the Ice Ages came to a close and these animals began to die out, those early hunters and gatherers began to focus more on smaller game and the wild plants around them. In time they would begin farming as well, tending fields of corn, wheat, rice, oats, and barley—crops derived from plants that were once grasses. In parts of Europe, Asia, and Africa, they took to tending herds of cattle, goats, and sheep, all domesticated relatives of the same animals their ancient ancestors had hunted before them. Over much of the New World, however, that herding way of life would not really take root until the arrival of the first Europeans more than 10,000 years after man's first arrival in the Americas. While the herds of cattle and sheep on the Great Plains brought little change and simply replaced the buffalo who had grazed there for thousands of years, the introduction of grazing animals into the deserts that lay beyond the Rockies would change the land almost completely. While the floors of valleys and basins in the Colorado Plateau and the neighboring Basin and Range were covered with grasslands, for several thousand years they had been grazed by nothing more than antelope, rabbit, and desert bighorn sheep—in effect, they were not grazed at all. That simple fact of history made them far different and far more susceptible to abuse than grasslands in other wetter regions that had evolved under the steady pressure of grazing. While the Anasazi planted fields of corn, beans, and squash, and built elaborate pueblos of masonry and adobe, the only domesticated animals they possessed were the turkeys and dogs. The Spanish would bring with them not only new culture, language, and religion, but new animals—herds of cattle, sheep, goats, and horses—that would eventually change the native landscape here as profoundly as the Spaniards themselves would change its native peoples. In this case, however, the changes would not be caused so much by the Spaniards themselves, but by the Americans who followed them.

In the late 1800s, ranchers and cowboys were drawn not only to rich, dry grasslands of southern Arizona, New Mexico, and Texas, but also to the rugged world of the Colorado Plateau as well. While miners discovered seams of silver and gold in the Rockies and Sierras, cattle and sheep thrived in the deserts between them during the cattle boom that followed the close of the Civil War. The canyons and deserts here, however, were—and, for that matter, still are—a difficult place for running livestock. Water was a constant problem and the herds had to be kept constantly on the move. Much of it was, as cowboys of the time liked to joke, 10-by-80 range. A steer, it was said, had to have a mouth ten feet wide and be able to run at eighty miles an hour to find enough food to survive. Down in the canyons cattle and sheep wandered off and were never seen again. Flash floods and river crossings claimed others as the animals were driven from place to place across the rivers and canyons of the Green, Colorado, San Rafael, and San Juan. Higher up there were problems as well. Grazing along the tops of dry mesas and buttes, the cattle would smell water in the canyons below and wander out onto inaccessible ledges of rock and become rim-rocked, unable to move forward or back. Some fell to their deaths. Others died of thirst in sight of running water, staring at the rivers and streams below. But just as the canyon walls concealed seeps and springs, basins and valleys and canyon floors also concealed hidden pockets of virgin grassland, areas of rich soil and water scattered across the desert like small green lakes. All one needed to turn these pools of grass into a magical source of cash was a herd of cattle or sheep. The desert grasslands here were a resource that had taken several thousand years to build as the result of a slow and steady accumulation of soil, nutrients, and life. In the space of just a few years, it would all but disappear.

In spite of the incurably romantic image of ranching and riding the range, the herds here on the Colorado Plateau in the late 1800s were not owned or controlled by small, independent operators trying to pull a ranch up by its bootstraps but by groups of wealthy investors from New York, London, and Edinburgh. The cowboys were little more than hired hands. From its headquarters near Moab, Utah, the Pittsburgh, Carlisle and Lacey Cattle Company ran more than fifty thousand head of cattle on a swath of land some sixty miles wide that reached southward from Moab all the way to Gallup, New Mexico. It covered more than six thousand square miles—almost all of it publicly owned land. During the annual cattle drive residents of nearby towns like Mancos, Colorado, claimed that it took three days for the company's herds to pass by. Others operating in the area, like the Aztec Land and Cattle Company, had holdings that spread over some 1600 square miles in a checkerboard pattern laced with farms and fields of Mormon pioneers. Fence lines and property lines, however, meant little to company officials, who simply ran their herds as if they held title to it all. Through a series of beatings and barn burnings they eventually drove the farmers and their families off the land—only to later sell it back to them under government pressure at eight dollars an acre—five to ten times the going rate of the day. The land here, however, was never rich and never would be. While the cattle companies made a killing, the boom was short-lived. Their herds were far larger than the land could ever support on a long-term basis, and they grazed it bare in a matter of decades. The deserts here were mined—not for gold or silver but for grass.

Heavy grazing damaged the land not only by erosion as hooves and feet broke up the ground and pounded fragile soils but also by cropping, as the animals stripped away palatable grasses and shrubs, leaving behind collections of nearly inedible weeds and poisonous plants. Diverse collections of grasses, forbs, and shrubs gave way to a monoculture of sagebrush and snakeweed. Elsewhere nonnative plants like cheatgrass moved in, far less nutritious and far more prone to fire than the grasses they replaced. Today well over half of the desert rangeland in the Intermountain West is what range management specialists term poor condition—capable of supporting only a fraction of its carrying capacity with respect to both livestock and wildlife. While they try to repair the mistakes of the past, ranchers and land managers in the region are stuck with the difficult task of learning how to live within the limits of the landscape.

In the late 1840s the Mormon scout Parley P. Pratt was sent south from Salt Lake City to look for possible sites for new settlements for the growing ranks of the church. His route took him along the high plateaus that mark the western edge of the Colorado Plateau. South of Nephi at the base of the Wasatch Plateau, he found what he thought was the prettiest grassland he had ever seen. In places it was so tall that Pratt and his men had to stand on the saddles of their horses to see over the top of the grass. While the grassland covered the curving edges of the valley as the land rose up to meet the sheer walls of the plateau, the center of the valley below was covered with big sagebrush—a sign of fertile soil.

Today, 150 years later, that pattern is nearly reversed. The thickets of big sagebrush that once carpeted the floor of the valley have long since been cleared away to make way for fields and farms while heavy grazing by cattle and sheep have transformed its grasslands above into tangles of sagebrush and scrub.

As we head toward Ephraim, Eldon Durant McArthur sniffs at the pungent smell of sage seeping in through the closed windows of the truck and turns to me and smiles. "I have to admit, I'm a sage-brush hugger." His words seem almost heretical for one whose ancestors were among the first Mormon pioneers to settle St. George, a hundred miles or so to the south. Undoubtedly they spent no small amount of time wrestling with sagebrush: clearing it off the valley floors and watching with dismay as it spread across the rangeland higher up. Professionally, however, his fondness for sagebrush is more understandable. Head of the U.S. Forest Service's Shrub Sciences Laboratory in Provo, Utah, and a geneticist, McArthur has spent much of his career studying desert plants like sagebrush, saltbush, rabbitbrush and bitterbrush in the Intermountain West. Unlike most ranchers and range management specialists in the region, McArthur does not view sagebrush as an unmitigated evil, but as a necessary and even valuable part of the landscape. The critical issue, he explains as we drive, is balance.

Overgrazing has allowed sagebrush to take over large areas of the West, expanding from scattered thickets and clumps to dominate entire basins and valleys. In its proper proportions, McArthur said, sagebrush is an important part of the landscape, its evergreen leaves providing forage when other desert and rangeland plants are dormant or dead—not a sole source of food, he added quickly, but a valuable supplement. While some varieties of sagebrush are unpalatable or unattractive to livestock, he said, others are quite nutritious. For decades ranchers and researchers in the West have looked for ways to eliminate sagebrush, but met with little success. Now, rather than looking for new ways of killing off sagebrush, researchers like McArthur have begun to look for ways of creating new ones—breeding new varieties that are palatable for livestock and wildlife and more resistant to fire as well.

Sagebrush is one of the most widespread plants in North America. It can be found all the way from British Columbia to Baja, California, and spreads across the western United States all the way from the Sierras to the Great Plains. It is perhaps the most common plant in the West. The most common variety is big sagebrush, but there are more than a dozen others: mountain sagebrush, silver sage-brush and dwarf sagebrush among them. While stands of big sagebrush can be found growing on the valley floors of the Intermountain West below 3,000 feet, others can be found high in the mountains, growing at 10,000 feet along the edge of a snowbank.

The piercing aromatic smell of sagebrush is almost characteristic of the West. After a rain it fills the air like a heady dose of perfume. Although its smell resembles that of the sage used in cooking, the sagebrush of western deserts and rangelands is totally unrelated—not a member of the *salvia* family like culinary sage, but a member of the wormwood family and distantly related to the sunflower. Seldom more than waist high, their branches do not rise up in straight leafy stalks like those of a sunflower, but are twisted and bent and covered with ribbons of fibrous bark and thousands of tiny green-gray leaves. While the sagebrush's branches are evergreen, the leaves change from season to season—small permanent ones that stay on the bush year-round and larger temporary ones that appear each spring with the first sign of rain and then fall off when the weather turns hot and dry. While warm desert plants like cacti protect themselves with thorns, these cold desert plants protect themselves with chemicals: not only those that inhibit the growth of other plants, but also those that ward off would-be grazers and browsers. The same terpene compounds that give them their characteristic smell can also give them a bitter taste—and in some cases even cause birth defects and tumors in the animals that feed on them as well—but the strength of these chemical defenses varies from variety to variety and place to place.

Not only do different varieties of sagebrush vary markedly, but individual plants seem to quickly adapt themselves to specific locations and environments as well. At a series of test plots on the flanks of Mount Nebo outside the small town of Ephraim, Utah, I spent a half hour or so walking around with McArthur looking at plantings of sagebrush. Even within plantings of the same variety of sagebrush, McArthur pointed out, there were differences in height and leaf shape based on where they were planted: high up on the slope, midway up it or down near the stream that ran near its base. Those differences seemed to hold true with their seeds as well, regardless of where they were planted. Those that had been taken from plants that grew near the base of the slope did poorly when planted up near its top, McArthur said—and vice versa. It was almost as if these simple shrubs had adapted themselves not just to a particular region or habitat—but a particular spot on it, as closely tied to the land as the soil beneath them.

Those signs of adaptation are not random or accidental. Sagebrush is able to evolve more quickly than many other plants by making use of a technique known as *polyploidy*, which makes them capable of doubling their chromosome number anywhere from two to eight times, giving them immediate genetic isolation. "Experiment is the key to evolution," McArthur said, and with the ability to create isolated populations almost immediately, sagebrush can do a lot of experimenting, creating specialized varieties and hybrids with subtle variations that make them uniquely adapted to a particular area or site—a particular level of sunlight or rainfall or even soil. Varieties of *Atriplex*, or saltbush, McArthur said, are even more adaptable, capable of doubling their chromosome number by anywhere from two to twenty times. It is no accident, perhaps, that saltbushes are one of the few types of plants that have been known to thrive on clay barrens scattered across the vast, dry badlands of the Mancos Shale that cover the Colorado Plateau. It is not just their toughness that enables them to survive, but their ability to rapidly evolve. Shaped by the extreme world of the deserts here, they have managed to carve out a place for themselves in the landscape.

That ability of native plants here to specialize and adapt affects the ecology of the region in subtle ways. In the 1950s, 60s and 70s, federal land management agencies in the West like the U.S. Forest Service and the Bureau of Land Management began a series of range improvement programs aimed at repairing the damage done to public rangelands by more than a century of over-grazing and abuse. Most focused on killing off unwanted plants like sagebrush and planting more productive grasses and shrubs. With regard to livestock, and in some cases wildlife, these plantings seemed to work quite well for a time, bringing noticeable increases in productivity. Most of the seedings, however, were done with non-native plants—grasses and forbs, for example, from Central Asia that seemed well adapted to both the dry landscape and grazing. Twenty to thirty years later, however, many of these plantings have begun to die off. The problem is not one of disease or drought, but long-term viability—at least on the Colorado Plateau and the neighboring Basin and Range. While these introduced species of plants were able to thrive at first, it now seems that they weren't well enough adapted to the peculiarities of the local environment to form stable populations. Rather than out-competing native plants, they seem to have been slowly overshadowed by them.

After a long period of abuse and decline, in fact, McArthur and other range scientists in the region believe that native plants on the Colorado Plateau are beginning to make a comeback in certain areas. That recovery, however, will not continue without careful management and thoughtful restrictions on grazing and other uses of the region. Instead of simply improvement, efforts to repair past damage to the region's grasslands and scrublands are now beginning to focus more and more on restoration—bringing back native species of plants.

With these trends in mind, researchers like McArthur have become increasingly interested in promoting the use of native plants and shrubs and looking for sources of seeds and ways to encourage their return. While early programs focused on range improvement, programs today are increasingly focused on range restoration—bringing back native species of plants and shrubs. McArthur's work with sagebrush is part of that program too—learning to work with native species of plants here rather than around them or over them. More than just idle curiosity, what drives this new approach to range management is a recognition of the special demands of the landscape here and the links of its native plants and animals to it. With careful management and thoughtful restrictions on grazing and other uses of the rangelands here native plants have a chance at making a comeback. What they cannot stand, however, is overuse. In other well-water parts of the United States native plants have been all but displaced by introduced species of plants—introduced varieties of weeds, grasses and trees. But while they have reshaped the landscape there, they seem to have hardly made an impact here. The combination of high altitude, low rainfall and thin soil has created an extreme environment where few outsiders can long survive. It is not just the evolution and history of the landscape here that is unique here, but the evolution of its plants and animals as well.

* * *

In spite of the intricate strategies for survival used by plants and animals here, the grasslands and scrublands of the Colorado Plateau are spare and often barren places, capable of supporting only a fraction of the life of the richer and greener worlds that lie beyond the Rockies and the Sierras. Out on the vast dry reaches of the San Juan Basin or the Painted Desert it can take several dozen acres of desert to support a single rabbit. In the clay barrens of central Utah that fringe the Henry Mountains, an acre of ground produces no more 100 pounds of organic material per year—compared to some 10,000 pounds per acre in prairie grasslands of the Great Plains and some 20,000 pounds per acre in the hardwood forests of the Appalachians. To put it in simpler terms, the productivity of two thousand square miles of desert equals no more than a single square mile of hardwood forest in the east—and that inescapable poverty of the landscape permeates aspects of life in the Four Corners region and the surrounding Colorado Plateau.

After an early spring rain, the land here is deceptively green. Visitors to the region are so taken by the exotic colors of the rocks and the blueness of the sky that they scarcely notice its barrenness, mistaking its broad reach for a sign of unlimited possibilities. People don't realize, Kimball Harper told me one afternoon at his office in Provo, that the margin for survival is paper thin. "I have relatives from southern California who come out here to visit," he said. "They hear me talking about population and the dangers of overpopulation and they say, 'What are you talking about? There's nobody out here. Look at all this empty space!' I say, 'You're welcome to it. Put down roots and try to survive there. You'll pack up and leave like everybody else has.'"

TWELVE

Anasazi

Doug Bowman's first job as an archaeologist was with the Ute Mountain Tribal Park at the base of Mesa Verde. One morning at 7:15 A.M. he got a frantic call from a Ute who had picked up a metate, a stone used for grinding corn and seeds, from the ruins of an Anasazi pueblo the day before without realizing what it was. "You've got to come out and pick this thing up," he told Bowman. "You've got to get it out of my house. I can't touch this thing. I haven't been able to sleep all night," he said. "He was like freako," Bowman recalled. "He was really bent." While the old ways are changing fast on the reservation, among traditional Ute the Anasazi and their ruins are still something to be feared rather than studied or admired. "They absolutely totally felt that they were bad spirits," Bowman said. "Ute religious leaders knew the Anasazi did a lot of burials. They could sense those spirits around the Anasazi dwellings. And their religion specifically said don't go near an Anasazi dwelling. Don't pick up one of their artifacts. If you do, it will contaminate your soul. You're going to have bad luck. You're going to die."

By the time the Ute reached the Four Corners area in the late 1400s, the Anasazi had been gone for more than a century. Although they were not ancestors or relatives, the Ute left the artifacts and pueblos of these ancient peoples alone, not so much out of respect but fear. Today the lands of the Ute Mountain Ute Reservation contain the best preserved pueblo ruins in the Southwest. While the ruins at Mesa Verde National Park are known worldwide, the best preserved ruins in the Southwest are found not within its boundaries but in the tribal land of the Ute that lies alongside it. Unlike the spectacular cliff dwellings at Mesa Verde, whose ruins have been shored up and rebuilt to accommodate a steadily growing stream of sightseers and tourists and whose most spectacular artifacts have been carted off to museums and collections worldwide, those on the lands of the Ute Mountain Ute have been left almost untouched, disturbed by nothing more than time. Although their relationship to the Anasazi is complex, today Ute guides escort small groups of visitors to ruins and archaeological sites on tribal land—offering a glimpse of the past that is little changed from that which greeted early explorers and archaeologists who traveled through the area one hundred years before.

Mesa Verde National Park draws more than seven hundred thousand visitors per year. Many see its ruins as the center of the Anasazi world. But while its well-known cliff dwellings are both beautiful and significant, "they are only the tip of the iceberg," Bowman said. "You have to think on a bigger scale." Traveling across the top of the mesa gives one only a glimpse of the Anasazi's ancient world. Its boundaries reach all the way to the edge of the horizon. "Think of the whole area as far as you can see in every direction. People drifting in and scattering about, but people everywhere. Down in the desert, up into the foothills and into the mountains. And from here to Nevada and down into Arizona. The Anasazi were everywhere. Eight hundred years ago there were probably more people living here in the Four Corners area than there are today."

In the 1960s Bowman was a hydrologist in southern California. Out in the field one day in Los Angeles County he came across a group of archaeologists working on an excavation. Curious, he wandered over and asked them what they were doing. "Excavating an Indian site," they told him.

"Where is it?" he asked.

"You're standing in the middle of it!" the crew foreman told him and then showed him a mano and a metate.

"That expanded my horizons a bit," Bowman said. At that point, he said, he became an amateur archaeologist. Before long he was spending his free time out in the deserts near Palmdale and Lancaster, wandering around and looking for things. A few years later he took a job for a firm preparing an environmental impact statement for proposed development of the oil-rich shales that lie beneath Uinta Basin in northeastern Utah. On the job he was regularly out in the field with archaeologists, showing them around the area and learning about the culture and history of people like the Fremont and the early hunters and gatherers who had inhabited the area before them.

When the oil-shale boom of the 1970s collapsed, he moved to Glenwood Springs, Colorado, not far from Aspen and took a job as an environmental coordinator for a large mining company. His interest in archaeology, however, was stronger than ever, and in his spare time he began taking classes in archaeology at Colorado Mountain College in town. Before long he had taken every course they offered and was looking for more. Teachers suggested he join the Colorado Archaeological Society to gain access to their classes and lectures. The nearest chapter, however, was nearly a hundred miles away in Grand Junction. Rather than commute, Bowman decided to start a new chapter and put an ad in the local paper announcing the first meeting of the Glenwood Springs chapter of the Colorado Archaeological Society. He expected no more than ten people to show up. Thirty-five attended the inaugural meeting. Bowman was elected chapter president.

Starting the chapter was like opening another door. With the group to draw on, he was able to get professors and professional archaeologists to come to town and give lectures and slide shows. The society also offered certification classes to become a licensed archaeological worker. "Then I was really hooked," Bowman said. Within a few years he had moved through the organization's ranks to become state president.

"Then at age fifty-five I decided I didn't want to be a hydrologist or an environmental coordinator anymore," Bowman said. "I wanted to be an archaeologist." He left his job in Glenwood Springs and enrolled at Fort Lewis College in Durango, Colorado, less then forty miles from Mesa Verde. He had his first job as an archaeologist with the Ute Mountain Tribe lined up before graduation. After three years with the tribe he moved to the Bureau of Land Management's

Anasazi Heritage Center in Dolores, Colorado. At the moment he is the director of the Cortez Colorado University Center, a museum and educational center in Cortez, Colorado, focused on the Anasazi and the native cultures of the Four Corners region. While Bowman has traveled in China, Korea, and Mexico, the focus of his archaeological interest is still the Southwest and the pueblo ruins of the Anasazi.

"The people who were here built this civilization over fifteen hundred years," he said. "The Four Corners region today has one of the highest concentrations of prehistoric sites of any place in the world. The average is 180 archaeological sites per square mile, and in some areas the number is as high as 230 sites per square mile." While that offers archaeologists a tremendous number of places to work, a site being anything from a five-hundred-room pueblo to a collection of pottery shards, even more intriguing is the relationship between the Anasazi and the cultures and peoples who surrounded them.

"In addition to what's available here, you can throw in four or five other states," Bowman said. "The Anasazi went all the way to Nevada. Throw in the Hohokam [a prehistoric pueblo people who inhabited much of southern Arizona] and you can take things all the way to Mexico." Cultures, crafts, and materials seem to have drifted northward as well. "We've got seashells coming in from the Gulf of California and the Gulf of Mexico," he said. "We've got macaw feathers from farther south coming up here and being used in sashes and ceremonial objects. It's amazing how interlinked North America was with central Mexico. And yet while they were linked to the Aztecs, Mayans, and Toltecs, those people never pushed directly in on them because they were separated from them by a vast desert. Corn, beans, and squash, the bow and arrow, ceramics, all came up from there. Some of the architectural skills probably came up from Mexico as well. So the tremendous influence of Toltec, Aztec, and Mayan culture is amazing. But it's thousands of miles away. It makes the Southwest a place of endless investigation."

Early investigators and archaeologists assumed that the prehistoric cultures of the Southwest had been transported wholesale from

Mexico. They left place names sprinkled across the ancient land-scape—Montezuma's Castle, Aztec Ruins—that suggested their strong ties to peoples and cultures farther south. Prehistoric pueblo people, they believed, had arrived in the Southwest only recently, perhaps no more than a few centuries ago. New finds coupled with new techniques of dating, however, soon revealed that both people and cultures here were far older than originally thought. Early man had not migrated into the area from Mexico a mere few hundred years ago but had been living in the region for more than ten thou-sand years. And while ancient people like the Anasazi had been clearly influenced by ideas and materials from Mexico, they also seemed to have developed in relative isolation—all part of a com-plex history of evolution.

After their arrival some fifteen thousand years ago, Paleolithic hunters and gatherers dominated the culture of the Southwest and early North America for several thousand years. By 5500 B.C., how-ever, the hunting and gathering lifestyle of Paleolithic times had given way to a more settled way of life known as the Archaic. As the climate warmed and giant animals like mammoths and mastodons became extinct, early desert hunters shifted to smaller game like rab-bits, dear, and mice, and became increasingly dependent on plants: the roots, leaves, seeds, and berries they could gather in the land around them.

Although this Archaic period lasted for more than five thousand years and represents an important transition between early hunting cultures and more modern farming ones, little more than the broadest details are known about the culture of Archaic times. In general the artifacts they left behind were less colorful or dramatic than those of either the Paleolithic hunters who preceded them or the Anasazi who came after them. Instead of the colorful and deli-cately shaped spearpoints of Paleolithic hunters, they left behind manos and metates that were used for grinding seeds and grains—a sign of their increasing dependence on plants. While the Anasazi built pueblos of masonry and adobe, the Archaic people who pre-ceded them lived in caves and brush huts and eventually pithouses.

Like the Paleolithic hunters who had migrated to the New World, traits of this Archaic culture were both widespread and worldwide. Woodland Indians in the eastern United States were as

likely to build pithouses as their relatives in the desert Southwest—
or for that matter the Archaic peoples of Europe, Asia, or Africa.
Differences in culture during these early times arose largely from differences in local resources—the plants and animals they depended
upon for survival. Over time those differences would become progressively more pronounced. Here in the desert, paradoxically enough,
agriculture would take root, and in time the cultivation of corn and
other crops would come to define both the history and culture of
native peoples here. In the traditional legends and stories of modern-
day pueblo peoples like the Hopi and Zuni, corn was a gift from the
gods—a magical source of life.

Scientists believe that corn was originally derived from tropical
grasses like teosinte that were domesticated in central Mexico sometime between 7000 and 5000 B.C. Passed from camp to camp and village to village, it would not reach what is now the southwestern
United States for nearly four thousand years. Although cobs and kernels of corn have been found in Archaic villages dating back to some
1500 B.C. in the western United States, archaeologists believe that
those early varieties of corn may have been used and cultivated much
like any other wild plant: Its seeds were scattered and then left to
take their chances—harvested if the crop happened to succeed and
forgotten if it happened to fail. Those early varieties of corn were
very different from those planted today. Although hardy, they were
far less productive than modern varieties of corn. Ears were only two
to three inches long and as thin as a pencil. One thousand years after
their arrival, Archaic peoples were still highly mobile, moving from
place to place in search of dependable supplies of food.

Literally and figuratively, however, with the arrival of corn the
seeds of a new way of life had been planted. In time, just like other
native plants in the desert, early varieties of corn would adapt themselves to the area and become a dependable supply of food. Soon they
would be followed by other crops from Mexico: squash and beans.
Slowly but surely, the tending of fields and crops would come to dominate the old hunting and gathering ways of life that had preceded it.
Clusters of pithouses slowly gave way to pueblos; baskets gave way to
pots—the beginnings of the elaborate world of the Anasazi.

The first signs of Anasazi culture would appear in the Four
Corners area around 100 B.C. as groups of these Archaic hunters and

gatherers began living in small villages or hamlets of pithouses near reliable sources of water and soils suitable for farming. Over the next few hundred years these simple pithouse villages would grow in size from clusters of two or three houses to villages of as many as fifty. The oldest of these early Anasazi villages at Mesa Verde dates to some 450 A.D. A few had what seemed to be large outdoor work areas and large kivas: circular underground chambers that, judging by both the traditions and practices of modern-day pueblo peoples and assemblages of artifacts found within them, were used for community and religious activities. Known as the Basketmaker People, these prehistoric pueblo peoples wove beautiful baskets and trays and stored their food in storage pits or cists in the floors of their homes and occasionally built others aboveground as well. Sometime during Basketmaker times the bow and arrow seem to have appeared, an addition to the spears, sticks, and nets that had been used by hunters in the past.

Around 700 A.D. these pithouse villages gave way to aboveground buildings of jacal——mud plastered over a framework of beams and sticks: the first true pueblos. Instead of being aboveground, however, the kivas of this early pueblo period remained underground like the pithouses from which they were possibly derived and became more elaborate, and many of what could perhaps be called the kiva's classical features appeared: stone benches and a ventilator shaft for bringing in fresh air and a firepit and siappu on the floor——the latter being a small hole in the floor symbolizing the entrance to the spirit world below, the earlier, simpler worlds through which man had emerged. Ladders led up through a central hole in the ground-level roof, the only point of entrance. Their size was typically as uniform as their style: from twelve to fifteen feet in diameter and some six feet deep. Larger great kivas appeared as well, but they were still far from common or characteristic. Pottery, seldom present during Basketmaker days, became more common, signaling another significant, or at least easily recognizable, change in culture, suggesting that life had become more stationary.

By 900 A.D. masonry and adobe had replaced the simple jacal structures of earlier pueblo times, and Anasazi villages became more widespread. While the culture of the Fremont farther north varied from place to place and canyon to canyon, the culture of the Anasazi was both widespread and uniform, with common styles of both architecture and pottery reaching all over the region, suggesting that trade and travel were prevalent throughout the area. Pueblos typically contained from ten to twenty small rooms, each room square or rectangular in shape and no more than eight or ten feet to a side. Blocks of rock and stone were used as bricks, carefully selected for size and shape, some pecked or finished for a neat appearance, with the spaces between them filled with mortar. Walls, both internal and external, were often plastered with adobe and sometimes painted or whitewashed as well. Doorways were small and low, less than four feet in height, low enough that even the Anasazi had to stoop when entering. Instead of hinged doors, they apparently covered the openings with skins or stone slabs in the winter. There was a distinctive layout to these pueblos as well: They tended to be linear, running from east to west, with their doorways and windows facing south. Storage rooms and storage pits were typically located on the north side of the pueblo, while its midden or trash pit was located to the south—almost immediately adjoining the rooms and work areas used for day-to-day living.

While the pueblos aboveground became more elaborate, so too did the kivas belowground. Great kivas became more widespread, growing in size and complexity. Some were more than fifty feet across. Posts and beams were replaced by heavy stone pilasters or columns for supporting the roofs. Benches and walls were lined or built of stone, and fire pits were made of masonry. Storage vaults and side entrances appeared in the sides of underground walls as well. Pottery became prevalent as ceramic ladles, bowls, canteens, and jugs appeared in a variety of colors and patterns: corrugated grayware, elegant smooth-sided pots covered with delicate glazes and colored with patterns of black and white, brown and red, many of them regionally distinctive and apparently traded between villages as well.

Perhaps even more striking than these changes in architecture and ceramics were the increasingly sophisticated means the Anasazi had developed for manipulating the region's scarce water supplies. Check dams, or small low terraces or walls for redirecting runoffs, canals, and retaining walls began appearing in pueblo fields. Rainfall was plentiful, and the Anasazi seemed to have reached a peak in both numbers and distribution during this early period. By 1050 A.D., small pueblos could be found almost anywhere in the Four Corners region, with similar styles of architecture, pottery, and farming.

By 1100 A.D. some small pueblos had begun to join together to form cities and towns. At places like Chaco Canyon in northwestern New Mexico's San Juan Basin, large, carefully planned pueblos of several hundred rooms appeared. Some grew by aggregation as village merged with village. Others were apparently planned and built from scratch. Outside their walls was an extensive system of roads, signal towers, and irrigation canals. The roadways are perhaps the most puzzling feature of all because the Anasazi had no carts or wheeled vehicles and the roads they built were far wider than necessary for foot travel, with principal routes some thirty feet wide and secondary ones fifteen feet wide. They traveled across the desert in straight lines. Rather than negotiate turns in a gentle curve, they made sharp right-angled turns. Steep grades were climbed by steps, while lowland areas prone to seasonal flooding and flash floods were crossed by causeways. Running outward from Chaco Canyon like spokes on a wheel, they seemed to connect its tight cluster of pueblos with villages and towns, some of them as far as eighty miles away. More than thirty-six separate villages have been discovered at Chaco Canyon. Estimates of Chaco Canyon's peak population vary anywhere from five thousand to twenty thousand.

Some one hundred miles to the north and one hundred years later, the Anasazi world would reach a peak at Mesa Verde, building elaborate pueblos of more than one hundred rooms and several stories perched on narrow ledges in the canyon walls. The same pattern would be repeated throughout the Four Corners area at places like Canyon de Chelly, Betakin, Keet Seel, and Hovenweep, the last a Ute word meaning "deserted valley."

In less than two hundred years, however, this sudden flowering of Anasazi culture in the Southwest would begin to wither and fade. By 1200 A.D. Chaco Canyon would be almost completely abandoned, and by 1275 Mesa Verde was being abandoned as well. After a thousand years of development and evolution, the world of the Anasazi here would come completely unraveled. By 1290 A.D. they would leave the Four Corners area forever.

$\bullet \quad \bullet \quad \bullet$

The dirt road runs past fields of dry grass and corn to a scrubby forest of pinyon and juniper that lies near the mesa's rim. It is late fall and the air is cool. Alongside the road the bright purple clumps of wild asters and yellow sunflowers that appeared with the late-summer rains are already beginning to wither and fade. As we park the car at the end of the road and begin walking cross-country through the forest, Sleeping Ute Mountain is almost directly in front of us to the south, its triangular shape rising up out of the desert like a pyramid. To the east is Mesa Verde, the sheer cliffs of its northern flank almost white in the late-morning light. The forest comes to an abrupt end at the side of a deep canyon. Looking south, it seems to cut into the edge of the mesa like a wedge, its converging walls tapering to a fine point that ends almost directly beneath our feet. From here there is only one way to follow the canyon: down. Twelve miles away and some two thousand feet below, its steadily dropping floor merges with the broader reach of McElmo Canyon at the base of Sleeping Ute Mountain. While springs are the only year-round source of water here in the side canyon, water flows year round in McElmo Canyon below. From its source near Mesa Verde, McElmo Creek twists and winds its way westward all the way to the San Juan River.

Low, broken walls of masonry fringe the canyon rim. Piles of square and rectangular blocks with pecked and chiseled surfaces lie half hidden beneath thickets of sagebrush, mountain mahogany, and serviceberry. Seven hundred and fifty years ago the canyon rim here was ringed by an Anasazi pueblo more than three times the size of the largest cliff dwellings found at nearby Mesa Verde. Known as Casa Negra, it contained more than four hundred rooms and fifty kivas, one of more than a half-dozen giant pueblos scattered among the side canyons and drainages that feed into McElmo Canyon near the Colorado-Utah line.

I am walking with Richard Wilshusen, the director of Crow Canyon Archaeological Center in Cortez, Colorado. A nonprofit research and educational institution, Crow Canyon is one of the few organizations still conducting long-term, ongoing archaeological research in the Four Corners region. From late spring until early fall, students and amateur archaeologists from around the world come to the center to learn about the culture of the Anasazi and work with the center's staff of archaeologists and scientists at ongoing excavations and research sites. Participants in the programs range from grade-school students to senior citizens. Next year they hope to start work on a series of excavations here at Casa Negra.

We spent several minutes looking for routes down into the canyon, looking for paths and handholds and picking our way across steep slopes and ledges. Pottery shards littered the ground in a rainbow of colors: black-and-white, red, and gray. As we walked, Wilshusen brought the ancient landscape alive, pointing out the tumbled piles of rocks from ancient room blocks and the lines of ancient walls. Circular depressions in the ground became giant kivas twenty and thirty feet in diameter. Halfway down the slope the bed of a dry creek appeared. Walking alongside it we could see the lines of terraces and check dams—low ridges and mounds built to direct runoff and rain—built by Anasazi farmers more than seven centuries before. Stopping near the side of a large boulder, I saw the broken shards of a large gray pot. Its surface was corrugated with ridges and bumps. Bending down to take a closer look, I could see the whorls of fingerprints pressed into the clay—marks of the potter who had once shaped this clay.

Although the pueblo here was far larger than those at Mesa Verde, these scattered walls, tumbled piles of rock, and broken shards of pottery are all that remain. Its condition is not atypical. The ruins here are not protected by overhanging walls of sandstone, nor have they been restored or rebuilt. Since the Anasazi departed, it has been left to the desert and the vagaries of time and vandals. Pothunters have long since made off with the more visible and spectacular remains that no doubt once covered the ground here—whole pitchers and pots, sandals, blankets, even ropes and charred ears of corn—but other details are still buried in the layers of sand and rubble that now fill its rooms and kivas: beams and posts that will enable archaeologists to date both its construction and abandonment; fragments of pottery and jewelry as well as traces of pollens and seeds and ash will offer some clue about the details of daily life and perhaps why this region was abandoned. Not only was this large pueblo built within sight of Mesa Verde, its people also walked away from the area at the same time, leaving their carefully constructed village to wander through the desert.

• • •

Just as the Colorado Plateau's colorful canyons and mesas became a center of geologic research, its well-preserved pueblo ruins would become a center of archaeological research as well. In the late 1800s geologists like John Wesley Powell, Grove Karl Gilbert, and Clarence Dutton would travel across the canyons and mesas of the Colorado Plateau and reshape the science of geology. In the early 1900s a similarly small handful of pioneering archaeologists would come to the area and reshape the science of archaeology as well. Building on the work of early investigators like Richard Wetherill, Jesse Fewkes, and Adolf Bandelier, they transformed archaeology from amateur collecting into a disciplined science, defining principles of excavation and study still in use today.

In New Mexico's Galisteo Basin, Nils Nelson would develop the principle of stratigraphic excavation. Borrowing the geologic principle that older deposits in undisturbed areas lie beneath younger ones, Nelson, who first published his ideas in 1914, found that one could trace the development of different pottery styles from ancient to recent by studying the successive layers of ancient trash deposits. A few years later, while working at Zuni, New Mexico, A. L. Kroeber would begin classifying archaeological sites by ceramic types, using changing assemblages of pottery styles and techniques to arrange them in time, in effect dating them, in much the same way a geologist uses sequences and assemblages of fossils to date an outcropping of rock. Then in 1917 Leslie Spier would refine Nelson's and Kroeber's ideas to develop a regional system of dating and classification for the ruins and archaeological sites of the Zuni–White Mountain area near the New Mexico–Arizona border. By the 1920s, archaeologists working in the area were no longer thinking in terms of the history of a single pueblo, but in terms of the history of an entire region. With dozens of scientists working on literally hundreds of sites, the flood of information was almost overwhelming.

In 1927 pioneering archaeologist Alfred V. Kidder invited a collection of Southwestern archaeologists to his field camp in New Mexico at Pecos Pueblo east of Santa Fe to develop a framework for organizing the increasingly detailed picture they were developing of prehistoric cultures in the Southwest. Focusing on the prehistoric pueblo people we know today as the Anasazi, they used changes in readily observable features like pottery and architecture to divide their history into eight steps or stages. The simple pithouse villages that had marked the beginnings of Anasazi culture in the Four Corners area were subdivided into stages of Basketmaker I, II, and III; the aboveground masonry pueblos that followed were subdivided into stages of Pueblo I, II, III, IV, and V. While the spectacular cliff dwellings of Mesa Verde and Canyon de Chelly were considered part of the Pueblo III stage, those that had been occupied in historic times were considered part of the so-called Pueblo V stage. With the exception of the Basketmaker I Stage, which is now considered part of the earlier Archaic, the Pecos Classification System developed by Kidder and his companions is still in use today. The meeting soon became a regular event, and in the years that followed other systems of classification were developed for other prehistoric peoples in the Southwest as well: the Hohokam of southern Arizona, the Mogollon of the Mogollon Rim, the Fremont of northern Utah and Colorado, and the Paytan of the lower Colorado River.

While study in the area is still firmly tied to these early roots and ideas, changes are afoot. "The field of archaeology is at a big crossroads," Wilshusen said. "Everything in the field has been cultural and historical up until now. If all you want to do is create a history, you can do it: how and when people settled an area, for example. Now we are thinking more in terms of what brings people together." Central to understanding these new ideas, Wilshusen explained, is understanding forces behind village aggregation, the growth of early towns and villages.

There is no question that agriculture changed the nature of prehistoric society in the Southwest, but archaeologists still debate the sequence of events that preceded its arrival—whether the demands of a growing population made its adoption essential for survival, or whether the appearance of agriculture caused the region's population to explode, forcing the ancient people here to change their way of life. The question is not an easy one to answer because people do not respond like a barometer in predictable and regular ways to changes in the world around them.

Whatever the true sequence of events, agriculture was widespread in the Southwest by 900 A.D. At about the same time, granaries and storage cists became prevalent, suggesting a noticeable increase in food storage. By 1000 A.D. population levels in the region were beginning to soar, and in less than two hundred years, water control features like check dams, canals, and terraces were wide-spread.

Once adopted, many archaeologists believe, the shift to agriculture would have been self-reinforcing. The demands of tending crops would tie people to a specific area for several months of the year, increasing the pressures created by hunting and gathering. As local populations of rabbit and deer began to dwindle and seeds and roots became harder to find, the need for planted crops like corn, beans, and squash would become even greater, tying the people even more firmly to a particular place and a particular way of life.

Other peoples and civilizations around the world, of course, went through the same changes and transformations as the Anasazi. Five thousand years ago the hunting and gathering lifestyle of Archaic times was found worldwide. As early peoples began to settle down, settlements, villages, and towns appeared, growing in size and complexity with time. In most other parts of the world, however, those traces of early life and transformation were followed by several thousand years of change and conquest, overprinted and all but erased by the successive civilizations and empires of ancient Europe, Asia, and Africa: the pharaohs of Egypt, the dynasties of China, and the empires of Greece and Rome. Here in the high deserts of the Four Corners, those ancient roots are still close to the surface and barely disturbed.

While the Anasazi were building their pueblos and cliff dwellings at Chaco Canyon and Mesa Verde, the Moors were building the Alhambra in Spain and the French were beginning the construction of the Cathedral of Notre Dame in Paris. As they began to abandon their pueblos and began migrating across the desert, Marco Polo had traveled to China and the Knights of the Fourth Crusade had sacked Constantinople. While wandering clans and peoples left their signs and symbols scattered across the cliffs and rocks of the desert, writers like Francis Bacon and Thomas Aquinas were at work and the first classes were being taught at Oxford and Cambridge. The juxtaposition of these facts should not be taken as a commentary on values or abilities, but rather a commentary on history: The fact that the so-called ancient world of the Anasazi is surprisingly recent. For the region's present-day pueblo people the history of the Anasazi is as relevant and important as the medieval history of England, Spain, and France is for those of us whose roots are more closely tied to Europe.

While the culture of the prehistoric pueblo peoples was clearly influenced by the culture of the Mayan and Aztec peoples of Mexico to the south, they never developed the same type of highly structured society. As the differences between the settled world of the pueblo peoples and their more mobile neighbors the Navajo and Ute so clearly illustrate, the evolution of culture and society is by no means predictable or certain—either in the Old World or the New. While the world's culture was surprisingly uniform during Paleolithic and Archaic times, the rise of distinctive cultures and civilizations would send the world's people and races off in sharply divergent paths. Perhaps there was some critical mass lacking in ancient North America, a density of people or ideas that elsewhere led to the rise of empires and armies in other parts of the world—or then again, perhaps those living here simply saw no need for them.

Whatever the reasons, fragments of that common early human history—the rise of villages and towns—are still readily visible here, often buried only a few inches below the ground. And it is those buried facts that may one day bring archaeology out of the past and into the present. The building of any culture or society involves a combination of compromise and cooperation. By understanding not only the forces that brought these ancient people together, but also those that drove them apart as well, we can quite possibly learn what joins and divides our own world. Was it the need for safety and security that drove these ancient peoples together, or was it the expression of some deeper human need? In a world that is increasingly divided by questions of class and color, those seem like questions that have as much relevance for our own time as for the past—questions we would do well to answer, or failing that, at least contemplate and consider. While archaeologists have left few stones unturned in the Southwest, it is the possibility of beginning to answer questions like these that keeps drawing workers like Wilshusen to the area.

"I guess it's that somewhere out there is the Galapagos Islands," he said as we worked our way out of the canyon and back to the car. "There's some find, some sense, that we're close to the answer that will explain things. It's a theory of evolution. The transition from Pueblo I to Pueblo II, from small camps to villages. Why did it happen? What creates society? History? Landscape? Tradition? What are you going to put your finger on? Somewhere out there is the Rosetta stone, the artifacts and the ideas that are going to help us understand all this."

We left the upper end of Sand Canyon shortly before nine in the morning, skirting the ruins of an ancient pueblo and heading off into the forest of pinyon and juniper beyond. The air was fresh and cool. Light breezes played through the low branches of the trees, but by noon the day was warm and still. The forest floor was dry and bare, the cedarlike scent of the trees mingled with the sharp smell of the dust.

Standing in the shade of the trees, you could hear the gnawing of bark beetles boring into the soft wood of the pines. Overhead a solitary raven circled the canyon rim, a black shape against the blue sky. In the long periods of silence between his dry rattling croaks, you could hear the soft swish of the wind whistling through his feathers.

Since early morning we have not seen or heard anyone outside of our own small group. Eight hundred years ago, however, the five-mile reach of Sand Canyon was home to several hundred Anasazi. At its peak, Sand Canyon Pueblo, whose ruins now ring the canyon head, had more than 450 rooms, 90 kivas, and a collection of towers and plazas. A few miles away at the canyon's mouth Castle Rock Pueblo had more than 60 rooms and perhaps as many as 15 kivas of its own. In between are more than a half dozen smaller pueblos, all part of what was once perhaps a single community. Today they are all quiet and still.

In the time of the Anasazi the upper reaches of the canyon here would have been filled with sound: the barking of dogs and the cries of children from the pueblo along with the softer voices of adults talking as they worked grinding corn or making string from yucca leaves. Middens would have been not just loose piles of dirt and broken pottery, but ripe hills of excrement and trash. On a warm summer's day the stench would have been perceptible from hundreds of yards away. The canyon and the pueblo here were not parkland or wilderness, but a place to live and work. Walking through the forests here we would have met dozens of Anasazi, some out tending small fields scattered across the top of the mesa, others gathering firewood or nuts and seeds.

For the past several years archaeologists and students from the Crow Canyon Archaeological Center have been studying and excavating the ruins of Sand Canyon Pueblo. That morning, however, we were not looking for the ruins of pueblos and kivas, but plants and trees. It was mid-September and I was out in the field with Mark Hovezak and three students from Crow Canyon's Environmental Archaeology Program for a day of plant collecting in the pinyon and juniper forests that surround Sand Canyon Pueblo.

Although archaeologists know a great deal about the construction of pueblos and kivas and the changing styles of Anasazi pottery, very little is clearly understood about the Anasazi's use of the landscape around them. It is only in the past decade or two that workers in the field have begun to regularly map the location of such things as agricultural fields and irrigation canals or to study the traces of pollens and seeds from the floors of ruins and surfaces of middens. While traces of pollens and seeds can provide a clue about the ancient environment and the wild plants gathered by the Anasazi, beyond the reach of the region's pueblo ruins are other sources of information about the past: not only the growth rings on trees, but also the plant collections contained in pack-rat middens, some of them more than one thousand years old, and the finely layered deposits in banks of arroyos and streams that reveal episodes of rapid erosion, flash floods, and droughts. Archaeologists today no longer simply dig up pots and catalogue their finds, but work in close collaboration with experts and scientists from a variety of fields: geologists, biologists, paleontologists, and others. Often as much work is done in the laboratory as in the field: studying shards of pottery under a microscope to identify distinctive types of temper or clay, or identifying traces of pollens and seeds to learn about the wild plants that grew near the pueblo and which may have been used by the Anasazi themselves.

Hovezak's work with the Ethnobotany Program is an attempt to take some of those ideas a few steps further, studying the ecology of the present-day forest to better understand what kinds of effects the Anasazi may have had upon it. We started that morning by carefully gathering up twigs and sticks from a variety of trees and bushes—pinyon, juniper, mountain mahogany, and squawbush—as part of a fuelwood study. The wood would eventually be burned back at the laboratory and the ashes analyzed in hopes of providing a baseline of information that could then be used to identify the types of charcoal and ash found in the firepits and rooms within the ruins themselves. The researchers had a number of questions in mind about the Anasazi's daily life: What kind of wood did they use to cook, make pots, or warm a kiva? Did they always use the same kind of wood, or did that wood change over time, suggesting that supplies were growing thin?

The goal is to understand not just when and where the Anasazi lived, but how and why. Over the summer, Hovezak shuttled back and forth between a number of test sites in the area: everything from garden plots planted and farmed in the traditional manner to estimate the productivity of Anasazi fields here, to studies of timber supplies in nearby forests to gauge how far the Anasazi would have had to travel to cut the poles and beams they needed to build their pueblos. Later that day we also gathered dead wood from a small test plot to determine the productivity of the forests here with respect to firewood and fuel. Although such work seems only remotely connected to archaeology, there were a number of critical questions behind it: how large of an area would the pueblo have needed to supply itself with fuel? What effect would the constant collecting of firewood have had on the forest? How quickly would they have used it all up and been forced to look elsewhere or even move? The questions are not simple and will not be answered in a few weeks of work. In all probability they will take several years of careful watching and collecting, but while the signs of these things are subtle, they are surprisingly basic to understanding the lives of the Anasazi.

By late afternoon we had worked our way back to the pueblo. While a group of students and staff members worked on excavating the ruins of pueblos and kivas, we spent the rest of the day looking for signs of seeds and fruits on prickly pear cacti, yucca, service-berry, pinyon pine, and more than a dozen other plants—all once harvested by the Anasazi. Fall was already well underway and most of the plants were spent, beginning to prepare themselves for the long winter ahead. On our hands and knees, we studied the tips of branches, twigs, and leaves for signs of continued growth, comparing the color, texture, and feel of different plants with those nearby. While these plants are all part of the canyon today, they were undoubtedly part of its past as well and Hovezak's perspective on them was geared toward that past: When were the wild plants the Anasazi depended upon alive and well? Which ones were available in the early spring when their stores of food had worn thin? Which ones lasted until well into the fall? What was the rhythm and pattern of life here?

Hovezak has visited these sites repeatedly over the year, carefully taking down notes and collecting samples, working with students and other staff members and sometimes alone, but the discipline and repetitiveness of the work seems to give him a certain pleasure. "Ethnobotany doesn't carry the same thrill of impending discovery as excavating a pueblo or kiva," he said. "It takes time for the results to make themselves apparent. Close study of the plants and animals here, however, helps bring the world into focus. Out in the clearings among the trees the sun was warm and bright. Amid stems and leaves of the yucca, servicberry, and clover, however, you could already see the signs of the approaching winter: Berries and leaves had already begun to wither and dry. Even then, however, preparations for the coming spring were underway. You could see it in the almost invisible buds and shoots that clung to the ends of branches and twigs, curled tight and fast, plans and designs for life in the coming year.

The landscape here is deceptive. On a warm fall afternoon, it seems gentle and rich and filled with endless possibilities, as if the soft warmth of the sun and the bright carpets of flowers—yellow sunflowers and purple asters—could continue indefinitely. Nights are cool and fresh, almost dreamlike.

In winter, however, the land here is cold and dead. Snow blankets the ground, and at times the temperature can drop to twenty below zero. In spring, winds laden with sand and grit blow at twenty to thirty miles per hour for days on end. In March the melting snow and the light spring rains bring water to the desert, but by late April the ground is already dry. By May the sky is blue and cloudless. The sun burns the ground like a torch. There is no rain or even clouds for several weeks—sometimes all the way into August. Summer thunderstorms are harsh and unpredictable, often washing away as much ground as they water. While creeks and arroyos beneath the track of the storm are filled to the brim by muddy floodwaters, a half mile away the ground can be completely dry.

That the Anasazi lived both so well and so long here is as much a tribute to their creativity as their pueblos and pottery. The margin for survival here, however, is paper thin, and the slightest change is capable of sending the whole world cartwheeling out of control. Against all odds and probabilities the Anasazi built up a civilization here based on the careful cultivation of crops. Agriculture here hinges on a delicate balance between rainfall and warmth: Move too high on the plateaus and mesas and the fall frosts will kill your plants long before they are ready for harvest. Move too low into the deserts and they will be burnt by the sun before they ever have a chance to grow. Even in the most promising locations here, rainfall is unpredictable, varying not only from year to year but from place to place and canyon to canyon. Survival depends on maintaining a delicate sense of balance.

The Anasazi overcame these difficulties by scattering their fields and gardens across the tops of mesas and along the floors of canyons in hopes that at least some of the plantings would be successful. Rather than large fields of a single crop, they favored small, carefully tended plots. They did not plant in rows but small mounds: half a dozen or more corn plants clustered together for support and protection from the wind. Beans were often planted right alongside, left to climb up the stalks of the corn like poles. They looked for the right spot for each crop: areas of good soil or swales and hollows where moisture collected, and south-facing slopes that were warmed by the sun in early spring. Check dams and terraces were put across the paths of small seasonal streams and drainages to collect rainwater and silt. Irrigation ditches carried water from streams to fields scattered across the wider reaches of canyon floors through a series of canals and ditches, complete with headgates and smaller temporary dams to regulate the flow of water. Fields were worked by stone hoes and pointed sticks. In places they mulched the soil with small pebbles and stones in square-meter plots, features still visible today more than seven hundred years later, appearing in aerial photographs of the region as checkerboard patterns of tiny squares. To store their hard-won harvests they built aboveground granaries and dug stone-lined storage pits or cists in the floors of their pueblos.

While farming represented a sharp break with the more limited hunting and gathering culture of the Archaic people they had evolved from, the plants and animals the Anasazi gathered from the land around them may have been as critical to their survival as their carefully cultivated fields of corn, beans, and squash. Artifacts found in abandoned ruins as well as the traces of bones, seeds, plants, and pollen from middens and work areas create a rich picture of Anasazi life. They ground the seeds of Indian ricegrass and nuts of pinyon pines into flour and used the flexible branches of willows to make baskets and trays. They used wild onions for seasoning and made a kind of lemonade from wolfberries. Chokecherries and serviceberries were gathered for food as well as acorns from gambel oaks and the padlike fruits of the prickly pear cacti. Other plants like yucca had literally hundreds of uses. Its roots were used to make soap, while its squash-sized seedpod was roasted and eaten like a potato. Bayonet-shaped leaves were split apart into threadlike fibers, then woven together to make string, rope, and cord that were, in turn, used to make string bags, hoists, harnesses, ropes, nets, sandals, snowshoes, and cradle-boards. Feathers were woven into strings and used to make down blankets, leggings, and coats.

Although the Anasazi domesticated crops such as corn, beans, squash, gourds, and cotton, the only domesticated animals they seem to have possessed were dogs and turkeys. While the dogs were perhaps used for work or hunting, the turkeys were valuable sources of both food and feathers and were kept in pens made out of daub-and-wattle fences within the pueblo. Other meat came from hunting: deer, antelope, and desert bighorn sheep were pursued on foot with bows and arrows and spears. Rabbits were herded into giant nets made out of yucca fibers, some as much as four feet high and dozens of feet long, and then killed with clubs.

Not only did the Anasazi hunt and gather like the prehistoric peoples who preceded them, some archaeologists have begun to believe that they may have moved seasonally as well, migrating up and down canyons and mesas to hunt and farm. While Doug Bowman was working at the Ute Mountain Tribal Park at the base of Mesa Verde, he plotted a series of paths and trails leading up from the floor of Mancos Canyon onto the top of the mesa, connecting the cliff dwellings and pueblos of the mesa top with those of the canyon below. "They knew each other. Obviously. From the top of the mesa to the bottom of the canyon," Bowman said. Some, he believes, may have even moved up and down the mesa with the changing seasons.

While the top of Mesa Verde is more than eight thousand feet high, the floor of Mancos Canyon reaches no higher than five thousand feet. "If you go up into Mesa Verde in January you'll see one or two feet of snow. Most of the vegetation is covered and buried. There's no animals. They all leave. It's an icebox," he said. "You drop down a thousand or fifteen hundred feet in elevation into the tribal park and a lot of the big pueblos are along south-facing cliffs in Mancos Canyon and there's no snow at all. The streams are still running. And where do all the deer go? They're all down in the bottomland because they've got to forage."

Rather than independent, self-sufficient communities, pueblos like those found at Mesa Verde may have been part of a larger community spread across several canyons. Similar ideas may also hold true for Sand Canyon as well. The smaller pueblos that line the canyon between the larger pueblos of Sand Canyon and Castle Rock may all be part of a dispersed community—smaller villages and settlements scattered around the periphery of a larger, central pueblo. Regionwide surveys of archaeological sites reveal that large pueblos like that found at the head of Sand Canyon may have been no more typical of Anasazi times than large cities like New York and Los Angeles are of the United States today. While such large pueblos undoubtedly played an important role in the region, smaller pueblos with ten or twenty rooms and only a handful of small kivas seem far more typical. Small villages may have served as satellites, enabling people to draw on a larger area for both farming and foraging, their dispersal guaranteeing that somewhere within the community reliable sources of food and supplies could be found if everyone worked together.

However dispersed and creative their use of the land, the farming and foraging of the Anasazi were not without their costs.

While we tend to think of prehistoric pueblos as stable and permanent, dates obtained by the studies of tree rings in the poles and beams used in the construction of pueblos and kivas suggest that the Anasazi moved quite frequently. While modern-day pueblos like Hopi Village of Walpi in Arizona and the pueblo of Acoma in New Mexico have been continuously occupied for more than seven hundred years, those of the Anasazi were typically occupied for only a generation or two, Bowman said—some thirty-five years for smaller pueblos and seventy to eighty years for larger ones. "Not too many were occupied continuously for longer than that, although we do find some where there was major rebuilding," he said. The pithouses of the early Anasazi were occupied for an even shorter time, no more than fifteen or twenty years, as if they had not completely abandoned their nomadic past.

"It makes sense that they would only be there for one generation," Bowman said. "After you've been living in one area for thirty-five years, you've really cleaned it out. You've used up all the resources. You've used up all the firewood. You've used up all the rabbits. You've used up all the yucca. You've got to walk two miles to find a yucca bush and halfway to Telluride to find a deer. They used up all the flora and fauna. And so finally they pack up and move to another canyon. They may only move ten miles. And very often the first thing they build is a pithouse. A pithouse is kind of like a mobile home or a camper, and then they start building a pueblo and with that pithouse they've got a hole in the ground and it turns into a kiva."

The transformation of the Anasazi dwellings from simple pithouses to several hundred rooms occurred in the Four Corners region over a period of some ten centuries: an endless circle of movements and migrations in which the region's canyons and mesas were continually discovered and rediscovered—settled and exhausted and then left to rest. After a thousand years it must have seemed like it would continue indefinitely. By the late 1200s, however, the fabric of life here would begin to unravel. In less than twenty years the Anasazi would leave the Four Corners region forever. Understanding how and why these ancient pueblo people left has been one of the most puzzling and persistent riddles in the Southwest. To those like the Hopi, Zuni, Acoma, and their brothers and sisters along the Rio Grande, however, it is no mystery at all, merely another step in their long history of migration and travel that have carried them up out of the First World, where time began, and onto the surface of our Fourth and present world—a part of their long search for the proper place and manner of living.

The junction of Sand Canyon and McElmo Canyon is rich and green. Irrigated fields of alfalfa and grass carpet the canyon floor, flanked by low hills of desert scrub. McElmo Creek flows through the canyon floor in a small canyon of its own, a twenty-foot-deep arroyo. Ranches and farms border the meandering path of the creek, shaded by groves of cottonwoods and thickets of Russian olives. On the north side of the creek the flanks of Sleeping Ute Mountain rise straight up toward the sky. A mile above, the forests of aspen that ring its summit are already beginning to turn, their leaves a bright blaze of gold. Down on the canyon floor the late-summer flowers are in full bloom, bright bursts of purple and yellow against walls of red and cream-colored rock. Merlins and kestrels skim across the fields, hunting for grasshoppers and mice. Mountain bluebirds sing from the tops of fence posts alongside the narrow road that leads east toward Cortez and Mesa Verde, their turquoise color a perfect match for the sky overhead.

In late September, a few days after my trip through the pinyon-juniper forests near Sand Canyon Pueblo, I am out in the field with Ricky Lightfoot and a group of students from Crow Canyon Archaeological Center watching the excavation of Castle Rock Pueblo, a counterpoint to the giant pueblo that lies some five miles up the canyon.

The pueblo sits atop a sloping ramp of bare rock that rises up out of the flat, green floor of McElmo Canyon toward the mouth of Sand Canyon above. Built around a solitary castle-shaped spire of rock more than sixty feet high, the tumbled walls of its rooms and kivas are still readily visible. At its peak the pueblo here was nearly as large as the largest cliff dwellings found at nearby Mesa Verde, containing perhaps as many as 60 rooms and 15 kivas. In places you can still see the broken walls of towers and rooms climbing through cracks and crevices in the sandstone. Photos taken by members of the Hayden Geological Survey who passed through the area in 1875 show rooms and towers sprawling across the top of the rock as well.

Tunnels and crawlways apparently joined the structures on top with rooms and kivas below. All that remains now are broken walls and piles of rubble.

The pueblo's location is intriguing and has led archaeologists to speculate about its purpose. Out on the sloping ramp of slickrock that borders the pueblo is an arc of giant boulders. The largest weigh more than half a ton. They were not carried there by flash floods or glaciers, but were put there by the Anasazi, as if they were intended to define the borders of a plaza or the entrance to the canyon above. Perhaps, some archaeologists believe, the pueblo here was a gateway for the larger Sand Canyon Pueblo above or a source of food and supplies. The ground here is some two thousand feet lower than that of the canyon rim above, and the growing season, as a result, is several weeks longer. Seven hundred years ago, when these ruins were occupied, McElmo Creek was not entrenched in a deep arroyo, but meandered across the canyon floor. While the flat land of its floodplain would have offered prime land for farming, the creek itself would have offered a year-round supply of water—a consideration the Anasazi would have been unlikely to overlook.

Lightfoot is wearing a dusty white straw hat and a green canvas vest whose pockets are stuffed full of pens and rulers. He squats on the circular floor of a partially excavated kiva, gathering up pieces of burnt wood and charcoal, remains of poles and beams from its cribbed, flat roof that will be sent off to a laboratory for dating.

The first archaeologists to excavate the great pueblos at Chaco Canyon some eighty miles from here in northwestern New Mexico burned the poles and beams they found in the ruins for firewood. Today the poles and beams from the Southwest's pueblo ruins can be dated not just to a particular decade or century but to a particular year. The key is not carbon 14 or some other radioisotope within the wood but the annual growth rings that mark its surface. Over the past sixty years the technique has been so refined that scientists working in the field of tree-ring dating, or dendrochronology, need a pattern of only twenty-five rings to assign an absolute date to ancient beams and posts.

The astronomer Andrew E. Douglass laid the groundwork for the modern science of dendrochronology in the late 1920s with his studies of tree-ring patterns in the Southwest. An astronomer with the Lowell Astronomical Observatory in Flagstaff, Arizona, Douglass was more interested in climate change than archaeological dating when he began his work. Reasoning that the growth rings of trees in the region should vary with rainfall, he hoped to track the Southwest's periodic droughts and fluctuating levels of rainfall by studying patterns of tree rings. While the concentric growth rings on the trunk of a living tree could be counted backward to determine the exact date, their relative thickness or thinness could be used as a relative gauge of rainfall. Years of abnormally high rainfall would leave relatively thick growth rings, and years of low rainfall would leave relatively thin growth rings; those during years of severe droughts would be almost imperceptible. Over time those yearly fluctuations in rainfall should leave distinctive patterns of tree rings behind, offering a glimpse of not only the timing of prehistoric droughts but their duration as well. By studying these patterns Douglass hoped to find some link between the Southwest's cyclical droughts and sunspot cycles on the sun's surface. While that link never materialized, his careful tracing backward from freshly cut trees through successively older pieces and fragments of wood (overlapping patterns enable one to move backward in time from living trees through the wood of successively older dead trees in a stairstep fashion) helped create a nearly absolute system of dating for prehistoric ruins in the Southwest.

While early archaeologists often excavated entire pueblos, hauling off boxes of pots and artifacts to museums and warehouses, research today is tightly controlled. Work here at Castle Rock has been under way for more than two years, but by the time the project is completed, they will have excavated less than 10 percent of the pueblo—much of it done in a series of one-by-one-meter or two-by-two-meter plots of buildings and structures visible from the surface and a few randomly selected ones as well.

In between work at his own site, Lightfoot travels from place to place at the dig, overseeing the work of a half dozen or so different students. Some are excavating a small section of the floor of a kiva. Others carefully remove the built-up dirt of several centuries that covers the surface of a plastered wall. The work is slow, a few inches per hour in some cases, the digging done entirely by hand. The location of each and every artifact is carefully noted: a broken piece of pottery, a charred piece of wood, the fire-cracked rocks of a burial hearth. "We do the excavating here with more care than is probably really necessary," Lightfoot said, "but we're working with people to try and educate them."

Federal laws requiring surveys and studies of archaeological sites in connection with the construction of dams, highways, power lines, and oil fields have spawned literally thousands of excavations in the Southwest. Purely academic research, however, is relatively rare and tightly regulated. Digs like this one at Castle Rock must be carefully planned and "designed," as archaeologists say. They must have specific goals in mind, specific questions they hope to answer before the go-ahead is received for excavations and studies on public lands—not only those managed by the National Park Service but also those managed by the National Forest Service and the Bureau of Land Management as well. A century of archaeology has left few stones unturned in the Southwest, and the goal of these restrictions is to make sure that a number of sites are preserved for the future when techniques and tools are even more sophisticated than they are today. With sites carefully parceled out, researchers like those at Crow Canyon are careful to glean every bit of information possible out of their sites. After spending the summer digging, reasearchers and scientists at the center will spend the long winter months inside analyzing their finds in the laboratory and writing up reports. "It's not so much what you find as what you find out," Lightfoot said.

There is no doubt as to the goal of research here at Castle Rock. Like Sand Canyon Pueblo above, the pueblo here was built in the late 1200s and reached its peak just as the Anasazi were beginning to abandon the Four Corners area. Researchers at Crow Canyon are hoping that the ruins here will help them understand the forces that eventually drove the Anasazi away. Tree-ring dates reveal that the construction of Castle Rock Pueblo began in 1261 A.D. and continued sporadically until 1275 A.D., when it was suddenly abandoned.

"It's a common misconception that abandonments are rare in Anasazi history," Lightfoot said. "But the fact of the matter is that they abandoned areas frequently. What happened in the late 1200s was different because they left and they never came back. Not only did they not come back, but others did not move into the area after they left."

People, Lightfoot said, have assumed that the Southwest was like Pompeii, that it was essentially abandoned overnight or quickly, but the truth of the matter seems to be quite different. Dates from a variety of pueblo ruins suggest that the Anasazi left the Four Corners area over a period of ten to twenty years, trickling out of the region pueblo by pueblo and clan by clan. In many cases, they seem to have known they were leaving as well. Beams and posts were pulled out of pueblos and kivas and carried off; so too were tools, pots, and supplies of food. Like a retreating army, they seem to have destroyed what they could not carry with them. Roofs and walls were pulled down and in some cases burned. Their concerns, Lightfoot suggested, may have been more spiritual than military— prompted by a need to keep outsiders from using or defiling their homes and burial grounds. Indeed, for the Anasazi, the two were often one and the same.

But while signs sugest their departure was premeditated, the reasons behind it are still hard to grasp. They left no written records behind them, and oral traditions among the present-day pueblo peoples offer only vague and enigmatic details (no one, for that matter, is even certain what type of language the Anasazi spoke, although it was undoubtedly related to at least one or more of the languages spoken by the modern-day pueblo peoples). The same tree rings that enable archaeologists to so accurately date the region's pueblo ruins also reveal that the area was hit by prolonged drought that began just before the area was abandoned and continued for several years after they had gone, lasting from 1270 to 1295 A.D. Detailed studies of other types of archaeological evidence suggest that a variety of other forces may have played a role as well: over-population, dwindling supplies of food and game, warfare, even religious prophecy. In the end, it may not be any one single problem that drove the Anasazi out of the Four Corners area but a combination of several.

"Droughts happen all the time around the world and they don't cause whole regions to be permanently abandoned," Lightfoot said. "And that's what is so curious about this one. Why didn't they come back? Why did everyone leave?" Data from tree-ring patterns and stream deposits suggests that major droughts have struck the Southwest fairly frequently, every 265 to 550 years—figures that seem to tie in with larger global trends. Not only have droughts been frequent in the region, but the Anasazi had also managed to survive droughts that were far more severe than the one that struck in the late 1200s. In fact, while the drought undoubtedly played a role in the abandonment of the Four Corners region, archaeologists like Lightfoot are beginning to believe, as he put it, that the "so-called great drought wasn't that great."

With the help of geologists and climatologists over the past decade or so, archaeologists have begun to look at more complex aspects of climate change in order to understand its impacts on prehistoric peoples like the Anasazi. Rather than thinking of climate change simply in terms of changing levels of rainfall, they have begun to think in terms of the rate of change as well, dividing changes in climate and environment into so-called high-frequency and low-frequency variations. High-frequency variations are those that typically change from year to year: sudden dry spells and cold spells. Low-frequency variations, in contrast, are typically spread over several years and even decades and include such things as changing rates of stream erosion and falling or rising water tables. While high-frequency events like a summer drought or a bitterly cold winter are readily apparent, they are often the easiest for people to cope with and are easily compensated for by stored supplies of food and firewood. Low-frequency changes, in contrast, are almost imperceptible with slight decreases in rainfall, for example, that last for years causing water tables to drop. As the groundwater levels drop, streams begin to cut down into their bed leaving fields alongside them high and dry. The problems build up slowly, but the results are catastrophic and often sudden.

With these kinds of ideas in mind, some archaeologists have come to believe that a series of moderately bad years may have been far more damaging to the Anasazi than a year or two of catastrophic drought, steadily depleting their carefully stored supplies of food and grain while the environment around them was eroding as well.

Therefore, while the so-called great drought that coincided with the Anasazi abandonment of the Four Corners area may not have been all that great, it struck at a time when the local environment was slowly deteriorating. At the same time population levels were not only high but highly concentrated. Clues are subtle but suggestive. While the bones of deer and elk are fairly common in middens of early pueblos, they become increasingly rarer with time, replaced by the bones of smaller animals like rabbits and mice, suggesting that food, or at least game, was increasingly harder to come by. At the same time, pollen samples from the floors and plazas of pueblos suggest that areas around them were being rapidly deforested as well—possibly as their needs for firewood and building materials grew. For the Anasazi, the changes taking place around them may have taken place so slowly that they saw no need for action until the steadily growing trend of disappointments had suddenly become a disaster.

At first it seemed like a piece of pottery or perhaps some plaster from an abode wall. By late afternoon, however, there was no doubt. What lay buried beneath the ground was not clay or stone but a human skull. The discovery triggered a chain of events: calls to the county coroner and the Bureau of Land Management. The next day a bone expert from Crow Canyon had come out to take a look, studying the patterns of teeth and bone, working the covering layers of dirt free with a split piece of pointed bamboo to avoid marking the bones. It lay on the ground upside down, its teeth and palate pointed straight up to the sky. The following day it would be completely reburied and the excavation shifted to another edge of the one-by-two-meter excavation pit in hopes of leaving both the skull and the bones that possibly lay near it undisturbed.

Archaeological research is a sensitive issue for many Native Americans in the Southwest, made even more so where bodies and graves are uncovered. While traditionalists among peoples like the Ute and the Navajo regard unearthing the remains of the dead as spiritually deadly, those among the pueblo people often see it, particularly when the remains of the Anasazi are concerned, as disturbing the bones of their ancestors. By 1997 all museums and archaeological collections in the United States will be required to return every piece of bone and their associated grave goods to tribal authorities. While the procedures and protocol for dealing with new graves and bodies in the field are still being worked out, Crow Canyon strives to be as conservative as possible—avoiding burials and bodies whenever possible and reburying the bones they uncover in the course of an excavation within a matter of days or even hours.

Almost anywhere else the discovery of the skull would have been surprising. While the Anasazi left pueblos and pictographs scattered throughout the Four Corners region, bodies and burials are relatively rare. In less than three years of work, however, archaeologists at Castle Rock have discovered the remains of twenty-three bodies—all, with the exception of one adult male, those of women and children—a number that seems even higher when one pauses to consider that less than 10 percent of the site has actually been excavated. This raises another disturbing possibility regarding the disappearance of the Anasazi—that at the time of abandonment the social fabric of the Four Corners area was beginning to come unraveled as well. The number of bodies, in fact, is not only high, but their end appears to have been violent as well: They sprawl on the floors of rooms and kivas as if they had dropped where they died or had been thrown through doorways and hatchways. Others are disarticulated—their bones separated and scattered over an area of several feet. Both the number of skeletons and the fact that they are almost exclusively those of women and children suggest that the victims here may have died in a raid, possibly while the men of the pueblo were out hunting.

The steadily rising body count at the site is disturbing. While the signs of Castle Rock's violent end seem obvious, it took archaeologists working there a long time to accept it, said Kristin Kuckleman, the Crow Canyon archaeologist who along with Lightfoot has been directing the excavations at Castle Rock. Hardest of all to understand were the scattered or disarticulated collections of bones they found mixed in with the collapsed roofs of buildings. "You wouldn't believe the kind of theories we came up with to account for the disarticulated remains in kiva roof-falls," Kuckleman said as we talked one afternoon in her office on Crow Canyon's campus outside of Cortez, Colorado. "We thought 'OK. They're building the roofs, but they don't have enough dirt to bury people so they're digging holes in the sides of the roofs and burying people in the roofs. And later when the roofs collapsed the bones sort of spilled out.' We came up with all sorts of bizarre theories," Kuckleman said. The signs of violence, however, were undeniable. With the startling finds at Castle Rock, workers at nearby Sand Canyon are beginning to rethink their own increasingly frequent encounters with bones.

At Castle Rock the suggestions of violence include not only the bodies that litter the pueblo ruins but the structure of the pueblo itself, the blocks of rooms that once climbed up its sides and stretched across its top. "I can't think of any other reason why somebody would want to build on top of a butte like that," Kuckleman said. "You had to haul up every rock. You had to haul up every bit of mortar. You had to haul up every bit of water to make the mortar. What kind of reason could you have for building on top of an almost inaccessible butte other than defense. There's no other reason to go up there. One structure would be a lookout. Four structures are not lookouts."

While the evidence of violence is extreme at Castle Rock, its defensive architecture is far from atypical. Pueblos from the late 1200s throughout the Four Corners region seem to have grown both increasingly larger and increasingly more inaccessible—clustered around the heads of springs or halfway up the vertical face of a sheer cliff. "You're really left with the inescapable conclusion that they are trying to protect themselves by building on those sites," Kuckleman said. "The fact that they're clustering around those springs really looks like they're trying to both protect themselves by being close together and protect some critical resource by just physically walling it off." However one might try to explain away the area's increasing aggregation—changes in agriculture, population, or religion—the possibility of struggle and conflict cannot be ignored.

While the Anasazi apparently fought no major wars or maintained no standing armies, their world was far from nonviolent. In historic times pueblo peoples fought regularly among themselves and with the Ute and Navajo who later came to live among them. Abandoned pueblos were often seen as casualties of war: villages and towns that had been destroyed, according to traditional stories, when they were attacked by others from nearby towns and pueblos. In the early 1700s the Hopi village of Awatowi was destroyed by other Hopis who objected to their conversion to Catholicism. The village and its church were completely destroyed. The men were executed, the women and children were taken away to other villages to be reeducated. On the way back, however, disagreements arose over the partitioning of captives, and most of the survivors were slaughtered—a story confirmed both by oral traditions among the Hopi and by archaeological finds. Fearful of being taken over by the Spanish, they felt forced to fight to survive.

Other disturbing signs of violence are found in prehistoric times among the Fremont and the Anasazi as well—some of them far more troubling than the number of bodies found at Castle Rock. At an excavation at Cottonwood Wash in southeastern Utah a few years back, Kuckleman and others found hundreds of pieces of human bone buried beneath the surface of a floor—the remains of perhaps three or four people, two adults and two children, splintered and broken into tiny pieces. Interpreting what happened, however, is a difficult task. Suggestions of violence anger many Native Americans, but the evidence seems irrefutable. "The more we find out about it, the more it becomes apparent that we don't have a good handle on it," Kuckleman said. "All you can say for sure is that something weird was going on." Violence is the Southwest's dirty little secret.

From time to time sensationalized accounts of these finds appear in the popular press, portraying the Anasazi as cannibals and igniting a firestorm of protest among both tribal authorities and archaeologists. Proving such claims of cannibalism is virtually impossible, but there is no denying that the signs of violence here can be disturbing and even terrifying. At their peak the Aztecs of Mexico offered daily sacrifices to keep the sun in motion. The Anasazi seem to have had their own need for sacrifice as well. Rather than evidence of war or cannibalism, they appear to be part of what archaeologists like Doug Bowman refer to as ritual killing.

The first signs of these rituals start appearing around 1050 A.D., small isolated killings of between five and ten people, with some containing as many as thirty. All seem to be killed in the same way. "They're not shot with a spear or an arrow," Bowman said. "They're bludgeoned to death, hit with a stone axe or something, inside of a building or inside of a kiva." The bodies are then butchered and stripped, every bit of skin, muscle, and tendon stripped from the bones. "The bones are then broken open and every single bone is separated from every other bone so that they're isolated," he added. "Modern-day pueblo peoples say that you have to do this to release the evil spirits." More than twenty of these sites have been found to date. Over in Utah archaeologists have also found evidence of facial scalpings as well—scalpings where the entire face was peeled away, the skin then painted and hung by a handle or cord, perhaps to be carried around or used in some kind of ceremony. Not only have actual scalps been found, but also depictions of them in pictographs and petroglyphs.

The killings are puzzling because on the surface Anasazi culture seems remarkably unified. "This was a culture in which everybody knew everybody. Everybody communicated with everybody. Everybody traded with everybody," Bowman said. Perhaps the killings were part of some Anasazi Inquisition, punishment of those who violated religious laws and taboos. Those who tried to leave the system became a threat. "That makes more sense than anything else," he said. "Someone starts doing something different, something that's considered weird and he upsets the whole balance of things. Everything they seemed to have done was in balance. Everything had to be done properly and in the old way. And suddenly some new guy brings in a skateboard and he blows the whole system," Bowman said. The cultures of modern Europe and America are not without their own episodes of persecution and unrest—stonings, crucifixions, and the burning of witches—and there is no reason to assume that the ancient world was any more immune to these violent obsessions than our own. "Everything points to internal and economic problems, and that something was going on in the late period. Whether people want to change the way of life we can't really say for sure. Maybe some people said no. You can't do that." Faced with a deteriorating climate and a changing social system, Anasazi society began to unravel. The center would no longer hold.

In spite of all these varied threads of evidence, the signs of social and environmental decay, to suggest that Anasazi may have struggled among themselves or with the landscape around them is not considered politic or polite by many academics and Native Americans. While the first Spanish conquistadors to reach the Southwest tried to transform its native peoples into Christians, we try to transform them into saints: perfect people who lived in total harmony with both the land and each other. Such views of history, however, are as idyllic as they are inaccurate. America before 1492 was neither a pristine wilderness nor unpopulated. Conflict and violence no more arrived in the New World with the first European than did culture and history.

The early peoples of North America used the land and life they found around them—and sometimes used it up completely. But with a population that was perhaps no more than 50 million and quite possibly less, in the New World there was always somewhere else to go, someplace else to move for a time while the land went about the slow and persistent business of restoring itself. Six hundred years after the Four Corners area was abandoned, the same logic that propelled the Anasazi to move from canyon to canyon in the space of a generation or two would be played out again by millions of European immigrants and pioneers who moved steadily westward clearing forests to make farms and plowing the prairies to make fields—and then learning how to live with the consequences: swirling dust storms and mud-laden streams. This time, however, the changes would be far more lasting and far more widespread than they had ever been in the past. The critical difference between these new arrivals and those who came before them that would have such as devastating impact on the land was not so much their increasingly powerful tools and technology—rifles, saws, and plows—as their rapidly increasing numbers.

Even the most careful among us do not live without having an impact on the land around us. As the desertification of Africa and the deforestation of places like Nepal and Tibet so graphically illustrate, subsistence economies can be every bit as destructive as our own industrial one. "We cannot refuse the exploitation of nature," the Mexican novelist Carlos Fuentes writes in his book *The Buried Mirror*. "It is the price of our survival." To accept this fact is not to excuse the excesses and abuses of our own society and time but to recognize the complexity of our own predicament. Returning a densely populated industrial nation like the United States to a subsistence economy would strip the continent bare in less than a century. The same tools and technology that have extended our reach and power have, in some sense, also increased our ability to use the landscape around us more efficiently and made it possible for our numbers to thrive and expand. Turning back the technological clock will not bring us back to some original state of grace.

While our offices, homes, and automobiles all too often isolate us from the natural world around us, we are no less dependent upon it than those who came before us. Beneath all our complexity, the water, food, and shelter that assure our own survival do not spring from our factories or the workings of our imagination but from the earth itself—the rhythm and flow of the changing seasons, the lives of plants and animals. In the end what saves us may not spring from our increasingly sophisticated manipulations of the world around us but from an increasingly sophisticated understanding of it: the possibility of finally reaching a sense of balance, understanding both the limits of ourselves and the world around us.

Seven hundred years after they were abandoned, the pueblo ruins of the Four Corners region still have much to teach us: Those who cannot resolve their differences seldom last for long. Those who cannot live within the limits of the land are often forced to leave it.

While the deterioration of both their land and their own society would drive the Anasazi from the Four Corners area, other forces were pulling them southward. In time they would rebuild their world at places like Hopi and Zuni and along the banks of the Rio Grande, learning from the mistakes of the past to fashion an elaborate and stable society that has survived right up to the present day.

It is a cold, clear winter's night in western New Mexico. The temperature has fallen well below freezing. The smoke from hundreds of fires hangs over the pueblo like a cloud, filling the air with the rich, sweet smell of burning juniper. Several thousand people line the streets that run along the southern edge of the pueblo: Zuni, Hopi, Acoma, Navajo, Apache, as well as Anglos from Albuquerque, Gallup, and Santa Fe. They are all waiting for the arrival of the Shalako, messengers from the gods who will bring the blessings of fertility, long life, prosperity, and happiness to the pueblo. Within the elaborate rituals and ceremonies of the Shalako is a reenactment of both the creation of the Zuni and their migration from a sacred lake village to their current "middle place," *Heptina*. The Zunis themselves call their pueblo *Hálona I'tiwana*, the Middle Anthill of the world.

I am standing up against a wall of the pueblo with a group of sisters and friends from St. Anthony's Zuni Indian Mission, a Catholic mission at the pueblo. We had walked down from the grounds of the school and convent several hours before, passing the time by visiting with friends and students and watching the crowd. The air is sharp and cold. I can feel it seeping through the thick layers of my wool coat and up through the soles of my boots. From time to time we stamp our feet to stay warm or move a few feet away to warm ourselves by the side of a large bonfire that burns in a nearby courtyard. Emerging from a neighboring apartment within the old pueblo, two men take a slowly cooked whole lamb out of an outdoor *horno*, or oven, that is shaped like a giant beehive. There is too much meat to carry by hand, so they toss it into a wheelbarrow and wheel it back to their apartment, leaving its rich, sweet smell wafting through the cold night air.

From across the river come the sounds of bells and chants. The Shalako and their attendants are approaching. Straining to see, we can make out their dark shapes walking across the wide fields that border the river—little more than a narrow stream by this time of year, December. As they cross a narrow footbridge, the black shapes take form, giant cone-shaped figures more than ten feet high, tapering upward to a birdlike head with rolling eyes and a cylindrical beak that claps and snaps, half a dozen in all. They pass quickly through the streets that run alongside the river, led by attendants from the pueblo's different kiva societies. The attendants chant prayers as they walk, their heads wrapped by headbands of knotted scarves, their shoulders draped with brightly colored Pendleton blankets. Bracelets and necklaces of silver and turquoise shine on chests and wrists. Zuni among the crowd push out into the street as the Shalako pass by, touching their sides and sprinkling them with offerings of cornmeal. After their appearance alongside the river the Shalako and their attendants spread out through the pueblo, heading off to the ceremonial houses that have been specially built for their arrival, where they will spend the night in dance and prayer, celebrating and feasting until the following afternoon. Police cars from the tribal police department lead the Shalako through the streets and the crowds with their lights flashing, passing out from the walls of the old pueblo and into the subdivisions of simple ranch-style homes and double-wides that surround the old pueblo—a mixture of old and new. In the wake of the Shalako the streets are filled with cars. The traffic will remain steady until early morning as the Zuni and their guests travel from Shalako house to Shalako house to see the messengers from the gods.

Over the past few years archaeologists and anthropologists seeking to understand the complex array of forces that may have pushed the Anasazi out of the Four Corners region have come to believe there may have been a strong cultural pull as well—the rise of the kachina cult. The sophisticated and colorful rituals of the kachina cult were all part of a new world-view, a new religion that bound the people, their gods, and the land together—the same force that leads the Zuni out into the streets of their pueblo to sprinkle offerings of cornmeal onto the shoulders of the Shalako.

To outsiders, kachinas are best-known as dolls or figurines. Among the pueblo people, however, the kachinas are also spirits, who bring both rain and blessings. Present in the traditional ceremonies of almost all present-day pueblos, kachinas vary from pueblo to pueblo and season to season. Those of the Shalako include not only the Shalako themselves but kachina spirits as well: the Long Horn with his solitary horn, the birdlike Salamobia with their whips of yucca, the Little Fire God with his painted body covered with spots and the clowning Mudheads with their masks of clay. The Shalako appear just once a year, a few weeks before the winter solstice when the nights are long and cold. Other months and other ceremonies bring other kachinas and other dancers—a collection of masked and painted figures that seem like a mixture of both man and animal—each with its own particular powers and blessings and its own particular season.

While portions of many kachina ceremonies and dances are open to outsiders, their significance and meaning are carefully guarded from outsiders. "According to legend," the anthropologist Bertha P. Dutton writes in her book *American Indians of the Southwest*, "in bygone days the kachinas used to come to the people when they were sad and lonely, and dance for them." They taught the people not only art but how to hunt and plant and raise their own food. "When rain was needed the kachinas would come and dance in the fields; then the rains always came." After a time, however, legends say that the people slowly began to lose their respect for the kachinas and even fought with some of them. Eventually they left and refused to return.

Instead they taught a few of the faithful young men the details of their ceremonies and showed them how to make masks and costumes, promising that as long as their instructions were followed and "their hearts were right," they would be permitted to act like kachinas. The kachinas themselves, in turn, would come to possess them, bringing rain and blessings as they had in the past, Dutton says.

The cult, then, in its rich and varied forms among the pueblos, provided a way for man not only to maintain contact with the gods but to bring the rain they so desperately needed to survive as well. In the precarious world of the Southwest, the kachina cult gave the pueblo people a means of control. Discipline was the key to survival: Everything must be done at its proper place and in its proper time. To perform the rituals properly and please the spirits, cooperation was needed from all within the village—the complex array of clans, societies, and priesthoods around which traditional pueblo society is built. The kachina cult not only helped assure the community's survival, but bound it together—requiring not only discipline and attention to detail but cooperation and unity as well.

While both the pueblos and the kachina cult would eventually reach their peak in the east along the Rio Grande, the first signs of the new way of life did not appear there but farther west, not far from present-day Zuni, in small pueblos and villages scattered along the upper reaches of the upper Little Colorado River near the New Mexico–Arizona line. The headwaters of the Little Colorado River lie within the high country of the Mogollon Rim that marks the southern edge of the Colorado Plateau, a green and fertile world where desert woodlands of pinyon and juniper give way to mountain forests of ponderosa pine and Douglas fir. At the turn of the century both Hopi and Zuni made pilgrimages to sacred lakes in the forests here, places where the kachinas are said to have first appeared.

Archaeological evidence suggests that the kachina cult began appearing here on the Mogollon Rim as early as 1275 while the abandonment of the Four Corners region to the north was still under way. Villages in the area suddenly began to dramatically increase in size. Distinctive styles of art and pottery appeared, many of them suddenly portraying the symbols and figures of kachinas: pictographs, petroglyphics, and elaborate murals on the walls of kivas. By the early 1300s important changes in architecture were appearing as well: While the round kivas of the Anasazi gave way to the square kivas of the kachina cult belowground, enclosed plazas began to appear in the pueblos aboveground—public places, as University of Arizona anthropologist and archaeologist E. Charles Adams suggests in his book *The Origin and Development of the Pueblo Katsina Cult*, where the kachinas could dance and perform their elaborate ceremonies for the people of the pueblo and their guests.

By 1325 the cult was firmly established in the area and had begun to spread to the Hopi and Zuni. As wandering clans and groups from the Four Corners area arrived, villages grew rapidly in size. Not only was the religious, political, and social power of the kachina cult able to bind these people together, it became even stronger with time. By the late 1300s the cult had begun to spread across the Painted Desert, reaching the Rio Grande in central New Mexico by the early 1400s. In a matter of years pictographs and petroglyphs of the kachina cult could be found all along the reach of the river all the way from Taos in northern New Mexico to El Paso in western Texas.

By the time the first Spaniards arrived, they would find the descendants of the Anasazi living in large pueblos scattered across the Southwest all the way from the Hopi mesas in Arizona to the banks of the Rio Grande in New Mexico; raising crops of corn, beans, and squash in carefully tended fields, their diverse families and clans linked by an elaborate cycle of rituals and ceremonies that bound them all together and sent clouds of rain drifting across the desert. Out of the ruins of the Four Corners region the pueblo peoples would build a strong and stable society that continues even today, capable of enduring not only decades of drought but centuries of domination and discrimination as well. That the pueblo peoples here were able to live so long and so well on the land had far more to do with togetherness than technology—the rise of a new way of thinking and a new way of life that emphasized the ties between both people and the land—a lesson that we are still struggling to learn today.

BOOK FOUR

East

PEOPLE AND THEIR cultures perish in isolation, but they are born or reborn in contact with other men and women, with men and women of another culture, another creed, another race. If we do not recognize our humanity in others, we shall not recognize it in ourselves.

CARLOS FUENTES
The Buried Mirror

THIRTEEN

A Ring of Fire

THE TOP OF Mount Taylor is more than two miles high. A solitary, cone-shaped peak seventy miles due west of Albuquerque, it overlooks the eastern edge of the Colorado Plateau. From its crest you can look east across a maze of forested mesas and hills toward Santa Fe and Albuquerque and see the high peaks of the Sangre de Cristos that mark the southern end of the Rocky Mountains. To the west, the view is entirely different. Looking northwest toward the Four Corners, you can see clear across the barren deserts of the San Juan Basin, all the way to the Chuska and Lukachukai mountains that straddle the border between New Mexico and Arizona—nearly a hundred miles away.

Other mountains lie to the south: the high peaks and plateaus of the Mogollon Rim in the distance and the low forested rise of the Zuni Mountains closer by. Like the higher and more jagged peaks of the Sangre de Cristos, the Zuni Mountains took shape during the Laramide orogeny some 50 to 70 million years ago, part of the same faulting and folding of rock that built up the Rocky Mountains. Like the Waterpocket Fold and the San Rafael Swell farther west, the uplift that created the Zuni Mountains bent the rocks here into a broad arch or bow. Unlike those to the west, however, the core of the uplift here was not soft sandstone or shale but hard rock from the continent's craton, or core—rocks more than a billion years old, the same age and appearance of those found on the floor of the Grand Canyon more than three hundred miles to the west. As erosion stripped away the softer layers above, these harder layers below were left behind as a line of low, rolling peaks. Today tilted cliffs of Navajo, Entrada, and Wingate sandstone parallel the sides of the mountain range, the flanks of a curving dome of rock that once arched over the tops of the mountains themselves.

Unlike these neighboring Zuni Mountains, however, Mount Taylor is not the product of faulting and folding but was built by eruptions and explosions of lava and volcanic ash that blew to the surface here between 2 and 4 million years ago. The Laramide orogeny that reshaped the Rocky Mountains and the Colorado Plateau, was followed by nearly 10 million years of quiet stability and steady erosion. Thirty-five-million years ago that period of calm ended with a bang as waves of volcanic activity began to roll through the Intermountain West, leaving flows of lava and volcanic peaks scattered all the way from the Sierras to the Rockies. In the Southwest those bursts of volcanic activity would all but encircle the Colorado Plateau, appearing in the San Juan Mountains to the north and along the edges of the high plateaus of the Wasatch Front to the west, the Mogollon Rim to the south, and the Rio Grande Rift to the east—enclosing the region with a ring of fire. The sizes of these explosions and flows of lava are unlike anything seen in the world today. The volcanic eruption that created the Valles Caldera northwest of Santa Fe in the Jemez Mountains was more than a hundred times the size of the recent eruption of Mount St. Helens in Washington State. It released enough lava to cover the continental United States with an inch-thick layer of volcanic rock, spewing clouds of ash that reached as far away as Iowa. The force of the volcano's explosion was so great that it left behind a craterlike caldera more than fourteen miles across. Other, older volcanoes to the south along the Mogollon Rim may have been even larger.

Active as this rift zone and volcanic fields were on the edge of the plateau, these late surges of volcanism in the West left the Colorado Plateau almost untouched. While eruptions and lava flows along the edges of the region built up solitary volcanic peaks like Mount Taylor in New Mexico and the San Francisco Peaks in Arizona, toward its center they barely broke the surface. They formed not volcanoes but laccoliths, as pockets of molten rock rose toward the surface like bubbles of hot tar, causing the earth's surface above them to blister and crack, creating ranges of high, solitary peaks like the Abajo, Henry, and La Sal mountains. Elsewhere jets of superheated steam shot through the earth's crust. Known as diatremes, they blew blocks of rock from the earth's mantle up through more than fifteen miles of solid rock all the way to the earth's surface. Today they rise up out of the desert like a collection of pillars and pipes: piles of altered rock from the earth's interior found at places like Shiprock, an eighteen-hundred-foot-high plug of volcanic rock in northwestern New Mexico, and the Hopi Buttes in northeastern Arizona.

Nowhere are the traces of that volcanic activity more apparent on the Colorado Plateau than here along its eastern edge near the hard black line of the Rio Grande Rift. The rift is not merely an isolated outpouring of volcanic rock but a vast seam in the earth's crust where the continent is actually being split apart. From its beginnings in central Colorado, the rift runs through New Mexico and continues southward into Mexico between the enfolding arms of the Sierra Madres before coming to an end a few hundred miles north of Mexico City. One hundred and eighty million years ago a similar rift appeared in the midst of the ancient supercontinent of Pangea to separate what is now North and South America from Europe and Asia to create the Atlantic Ocean. In the space of a few million years the Rio Grande Rift may split the continent again.

Farther north in Colorado the eastern edge of the Colorado Plateau is irregular and hard to define, lost in a subtle shift from mountain to mesa as the edge of the Rockies weaves in and out of the plateau country. Here in New Mexico, however, the edge of the Colorado Plateau is as plain and arbitrary as a line on a map. South of Albuquerque the edge of the Colorado Plateau is defined by the trace of the Rio Grande Rift. To the east is the broad swell of the Great Plains, an endless sea of grass reaching more than halfway to the Mississippi. To the west is the Colorado Plateau, a stairstepped landscape of deep canyons and high plateaus that reaches all the way to the Wasatch Front in central Utah. Many of the same layers of rock that run through the Colorado Plateau stretch eastward for hundreds of miles into the Great Plains as well: gray layers of

Mancos Shale and the brightly colored sandstones and clays of the Chinle Formation. Buried beneath an apron of sand and silt eroded from the Rocky Mountains, they break through to the surface only occasionally.

While volcanic peaks like the Manzano Mountains near Belen and the Magdalena Mountains near Socorro flank the sides of the Rio Grande Rift to the east, the profiles and shapes of other volcanoes are visible to the south: the Datil and Mangas mountains. Hot springs and thermal pools suggest that the pools of molten rock that fed these rifts and volcanoes are not all that far away. While Mount Taylor has been quiet for more than 2 million years, flows of lava in the valley below are less than a thousand years old, part of a sheet of volcanic rock known as *El Malpais*, literally "The Bad Country," that carpets the floor of the desert between Cebollita Mesa to the south and the nearby Zuni Mountains. From the top of Mount Taylor the lava flows of El Malpais seem to spread across the desert like a deep black pool. Not far from its edges are the pueblos of Zuni and Acoma.

After the Anasazi abandoned the Four Corners region, the Rio Grande seemed to draw them like a magnet. While the land near the river offered fertile, well-watered ground for crops, in the hills and mesas above it were rich forests of pinyon and juniper that were almost identical to those they had left behind in the Four Corners region little more than a century before. Close on the heels of the pueblo peoples, however, others were arriving here as well: roving bands of Navajo and Apache drifting southward from Canada and Spanish soldiers and priests pushing northward from Mexico, following the natural corridor of the rift and the river.

Just as the rift marked a sharp break in the geologic evolution of the plateau, so too did the arrival of these outsiders mark a sharp break in the cultural evolution of the region. After the close of the Ice Ages, a rising sea level had separated the Old World from the New for more than ten thousand years. With the arrival of Spanish soldiers and priests, the diverging paths of history and culture that had shaped these separate worlds would suddenly cross with results that were often both painful and deadly. Out of this conflict, however, a new culture and a new way of life would emerge. For better or worse, native culture and Spanish culture would become so intertwined over much of the Southwest that they would prove almost impossible to separate.

If you had been able to sit atop of Mount Taylor fifteen thousand years ago at the close of the Ice Ages, you would have seen bands of Paleolithic hunters stalking herds of mammoths and mastodons in the plains below. Looking east toward the Sangre de Cristo Mountains, you could have watched the glaciers that once covered their flanks slowly disappear. As the climate continued to warm, forests of pinyon and juniper below would begin climbing the flanks of the mountains and plateaus. In their place, desert grasses and shrubs from farther south would begin moving in: thickets of scrub and cacti and bunch-grasses. In time they would all but fill the flat reach of the valleys below.

People would come and go as well, bands of hunters and gath-erers traveling through the desert. Later small villages would appear—clusters of pithouses and small pueblos surrounded by crudely planted fields. From time to time eruptions and explosions of volcanic rock would send rivers of fire and molten rock coursing across the ground below. By 1300 wandering clans of Anasazi would pass by, heading south and east toward the Rio Grande and to nearby pueblos like Laguna, Acoma, and Zuni. Pueblos of five hun-dred and a thousand rooms would spring up in the desert, fringed by cultivated fields of corn, beans, and squash—far larger than any-thing that had ever been built before. Shortly after their arrival roving bands of Navajo and Apache would appear, living in small mobile camps scattered outside the more settled world of the pueblos, scattered across the deserts and plateaus below. By the late 1500s you would see caravans of Spanish soldiers, settlers, and priests pushing northward along the Rio Grande, their herds of cattle and horses raising up clouds of dust as they passed by—the future coming face-to-face with the past.

F O U R T E E N

Rifts and Volcanoes

T HE TRAIL HEADS out toward the lava through meadows of low green grass. It is mid-September, but after three days of rain, the desert here has come unexpectedly back to life. Wildflowers paint the fields of grass with bright splotches of color: flashes of orange, yellow, purple, and red from globe mallow, desert marigolds, four-o'clocks, and Indian paintbrush. After a half mile of open, easy walking, the meadow gives way to black rock, a river of basalt that runs across the ground for more than twenty miles. The trail does not falter but plunges straight ahead, its path across the lava fields marked by cairns of black rock.

To the north the low triangular shape of Mount Taylor reaches up into the turquoise blue of the sky. To the east, across several miles of broken lava, are the low, tan cliffs of Cebollita Mesa. The Spanish called this lava-filled valley near the base of Mount Taylor *El Malpais*, the bad country. A land of broken black rock almost impassable on horseback, much less in a wagon or cart, it was a place to be avoided, an obstacle to traverse in the dry desert valleys that ran west from Santa Fe and Albuquerque to the western pueblos of Acoma, Zuni, and Hopi. Before the arrival of the Spaniards, the pueblo peoples had been traveling across the volcanic badlands of El Malpais for centuries. Cairns marking the path here are several centuries old, part of a traditional trade and ceremonial route connecting the pueblos of Zuni and Acoma. Here and there one can see small bridges of piled-up rocks spanning deep cracks and gaps in the lava, built by the Anasazi. For a time they seem to have camped or even lived here among the lava fields, building clusters of ring-shaped structures out of blocks of basalt. Their design suggests that they may have been built by people from the great pueblos of Chaco Canyon that flourished between 900 and 1100 A.D. some one hundred miles to the north, or at least by people who were strongly influenced by them. In all probability they were living here when the last eruption of lava took place some seven hundred years ago.

Traditional stories and myths among the pueblo people say that the lava is the blood of a kachina who had been blinded by his sons for gambling. When they plucked out his eyes, the legends say, streams of fiery lava poured forth from his eye sockets. As these rivers of "fire-rock" moved across the ground, they destroyed everything in their path and sent heat waves curling up into the sky. When a curious raven flew too close, his feathers turned the color of charcoal—the black color they still bear today.

The lava fields here stretch from the edge of the Zuni Mountains to Cebollita Mesa, covering an area of nearly six hundred square miles. They did not come from Mount Taylor to the north but exploded up out of the ground from a network of some thirty volcanoes and a collection of more than eighty spatter cones and vents. There were at least five major flows of lava. The oldest cooled some 3 million years ago. The youngest moved across the ground here around 1100 A.D.—only a few hundred years before the arrival of the Spaniards. In places you can see baked and burnt layers of soil where the lava flowed over open ground, burning the plants, trees, and dirt beneath it. Near the edge of the Zuni Mountains, lavas less than ten thousand years old lie alongside rocks more than one billion years old, a collection of granites and gneisses from the continent's stable craton.

The black rocks here, however, are not part of the craton but pieces of the fluid mantle below. They came up out of the ground at more than a thousand degrees Fahrenheit, so hot that the rocks were actually fluid, capable of flowing across the ground like molten steel from a blast furnace. Some slid across the ground in sheets. Others were so laden with gas that they exploded, shooting molten chunks of rock into the air—volcanic bombs—that hardened before even touching the ground, sailing through the air for hundreds of yards. Elsewhere are conical cinder cones, volcanic vents surrounded by sloping, circular aprons of volcanic cinders—pebble- and marble-sized bits of cooled lava so shot through with tiny pockets and holes left by bubbles and gas and air that they feel as light as pieces of sponge or foam.

Although the ground here is covered with rock, it is far from bare. Climb up to the top of a ridge of volcanic rock or a low cone of basalt and you can hear the blocks of lava rub and grind beneath your feet with the hollow sound of glass on glass. On top you can look out across a mixture of black rock and green trees. Forests of pine and fir dot the surface of all but the youngest flows. Some are dry forests of pinyon and juniper like those found on the top of nearby plateaus. Others are damp forests of ponderosa pine and Douglas fir like those found in the high forests of Mount Taylor or the neighboring Zuni Mountains. While the initial surge of the molten lava killed everything in its path, once cooled it provided protection for delicate roots and helped trap moisture and rain-water. Nothing goes to waste here in the desert. Even where the ground seems to be paved with boulders and stones, crusts of lichens and moss color the rocks here with shades of orange, yellow, and pale green. In hollows where thin soils and leaves have collected, you can see the tracks of deer. In the space of a few thousand years, forests have been able to gain a foothold here on the lava.

Younger flows, however, are still black and bare. Farther east the lava is so fresh and glossy that it looks like it had cooled only a fews day before. While the rocks within a single flow are surprisingly uniform with respect to their chemical composition, the irregularities of cooling and motion have left them with strikingly different appearances. In places they seem to have cooled into angular and irregular blocks—almost like pack ice in the Arctic, formed by the repeated freezing and thawing of ice. Elsewhere the lava has a soft, almost ropy look to it, like a flow of warm tar. Those block flows of lava are known as *aa* (pronounced "ah-ah"), while the ropy flows are known as *pahoehoe* ("pa-hoy-hoy"). The words are Hawaiian, used by natives on the Hawaiian Islands to describe the texture of lava flows from the islands' volcanoes and now widely used by scientists. Pahoehoe lava forms when the lava begins to thicken and cool. Aa lava forms when a hard crust of cooled lava is broken into slabs and blocks by the still-fluid lava moving beneath it.

On the flanks of Hawaii's Mauna Loa, basalts are capable of moving across flat ground faster than a man can walk—at some five miles per hour. As the lava flows, however, it cools from the outside in, forming a chamber or conduit of molten lava below. In places here at El Malpais that still-fluid lava below managed to flow completely away, leaving its surface scored by a series of tunnels and tubes. One near Bandera Crater on the western edge of the lava field is more than fifteen miles long, riddled with caves and collapsed sections of roof, marked at the surface by what appear to be a network of gullies and ditches. Another a few miles away, known as Dripping Lava Cave, is more than a thousand feet long with a ceiling thirty to seventy-five feet high. Some of these caves and tubes are floored by permanent pools of ice. Cold air sinks into the caves from late fall to early spring. Protected by the thick, insulating layers of lava, the interior of the cave heats up only slightly during the hot months of summer. In the midst of an ancient river of fire you can find sheets of ancient ice.

The eastern edge of the Colorado Plateau swings out in a wide arc from the Uinta Mountains to enclose the Piceance Basin of northwestern Colorado. A sagebrush-covered desert cut by the White and Yampa rivers as they head toward their junction with the Green River near Dinosaur National Monument in Utah, it seems in places as barren and open as New Mexico's San Juan Basin to the south. Here in the northeast, the surface of the Colorado Plateau is broken by a series of broad mesas and plateaus: Battlement and Grand mesas, the Roan and Uncompahgre plateaus—the latter a persistent reach of high ground that first appeared as part of the Ancestral Rockies some 300 million years ago.

South of the Uncompahgre Plateau the edge of the Colorado Plateau curves westward around the San Juan Mountains, a mountainous fist of land that punches outward into southwestern Colorado, reaching almost all the way to the Four Corners. Part of the Rockies, the San Juans first appeared during the Laramide orogeny some 70 to 50 million years ago. Later they were built up even higher by a series of widespread explosions of volcanic lava and ash. The timing of those eruptions, between 35 and 25 million years ago, roughly coincided with both the appearance of the Rio Grande Rift to the east and the eruption of volcanoes and lava fields along the Mogollon Rim to the south. While the Laramide orogeny that started the rise of the Rocky Mountains, would lace the surface of the Colorado Plateau with faults and folds, the widespread volcanism that followed it would define the edges of the region even further.

Where the plateau skirts the edge of the San Juans its surface is covered with forests of pinyon and juniper and deserts of grass and sage. Near Mesa Verde the edge of the Colorado Plateau swings eastward again, paralleling the flanks of the San Juans. In the La Plata Mountains just east of Mesa Verde you can see benches of volcanic rock at the base of the peak. Cooling has left their faces lined with columnar joints of basalt that look almost like pillars—so large they are visible from ten to twenty miles away.

Farther east past Mancos and Durango, high peaks seem to surround the small town of Pagosa Springs on every side as the Rockies swing southward again and head into New Mexico. Following U.S. Highway 84 south toward Chama and Tierra Amarilla, you can see how the mountains begin to break up into plateaus and benches, an almost steplike progression from mountain to mesa. By the time you reach Abiquiu some eighty miles south of the border, the transition seems almost complete. High cliffs of Chinle, Entrada, Navajo, and Wingate sandstone rise up right from the highway's edge, the same brightly colored rocks that grace the Waterpocket Fold and the canyon country of southeastern Utah. Here they form a backdrop for Ghost Ranch, where Georgia O'Keeffe came to live and paint, creating her stark and surreal images of the Southwest.

Below Abiquiu the edge of the plateau swings east to include the Jemez Mountains behind Los Alamos. At their crest is the fourteen-mile-wide Valles Caldera, created by a volcanic explosion only a few million years ago. Near Espanola the edge of the Colorado Plateau merges with the Rio Grande Rift, a wide valley whose edges have been defined by faulting and a series of volcanic eruptions. Farther north toward Taos the rift appears as a narrow sage-covered valley at the foot of the Sangre de Cristo Mountains, its edges fluted by fans of sediment that spill out from the mouths of canyons. Its surface seems to curve like a shallow trough, dotted by the distinctive cone-like shape of small volcanoes and cinder cones. Near its center, however, the Rio Grande Rift tumbles through a 650-foot-deep canyon between walls of hard black lava. This is no simple canyon formed by the steady wearing away of rock but a deep tear in the earth's crust formed by flows of basalt—hard dark flows of lavas from the earth's interior. Here on the edge of the Colorado Plateau, the continent is trying to tear itself in two.

The Rio Grande Rift begins as a narrow sliver in the Arkansas River Valley of central Colorado and continues southward into the San Luis Valley. Shaped like a spindle, more than one hundred miles long and fifty miles wide, it separates the San Juans from the Sangre de Cristos, dividing the southern Rocky Mountains in two. Approaching Taos, the trace of the rift narrows again as the mountains draw together. Its southern end is dotted with hundreds of small volcanic cones, reminders of the rifting and volcanic activity that gave the valley its shape. By the time it reaches Albuquerque the Rio Grande Rift is more than thirty miles wide, and it becomes steadily wider as it heads south, all the way to León or Guanajuato at the northern end of the Valley of Mexico.

Rifting seems to have begun some 30 million years ago along the course of the Rio Grande, but there is no clear consensus of opinion among geologists as to the forces responsible for its appearance. Its timing coincides not only with a flood of volcanic activity around the margins of the Colorado Plateau but also with the beginning of rifting farther west along the East Pacific Rise, which would ultimately split Baja California from the Mexican mainland.

One hundred and fifty million years ago a rift zone much like the Rio Grande separated North and South America from Europe and Africa to create the Atlantic Ocean. Rifting not only opened the Atlantic but drove the continent westward through the Pacific. Thirty million years ago, that western edge of the continent ran right over the top of the East Pacific Rise. The collision was not head-on but angular. While portions of the rise still survived off the coasts of Washington and Oregon, farther south the rise began to try and work its way into the continent, creating a narrow valley or depression, stretching northward from Mazatlán and into southern California. Five million years ago rifting began in earnest along the west coast, severing Baja from the Mexican mainland to create the Gulf of California. At the northern end of the Gulf the East Pacific Rise gives way to the San Andreas Fault that runs northward through coastal California all the way to San Francisco. Above San Francisco, the fault heads out to sea, paralleling the coast as it heads north until it merges with another spreading rise, the Gorda Ridge near Cape Mendocino, not far from the Oregon border. Movement along the fault is not pushing coastal California from the mainland but slowly driving it northward.

There is no corresponding San Andreas Fault for the Rio Grande Rift, but its similar timing is intriguing. The slip and slide of continent and seafloor that occur along the San Andreas Fault have broken parts of both the California and Oregon coasts into blocks and caused them to spin counterclockwise, some of them by almost as much as ninety degrees—almost like ball bearings. Some geologists have suggested that the Colorado Plateau may be undergoing a slight rotation of its own. If that is true, the Rio Grande Rift may have taken shape as a tear in the earth's crust as the solid block of the plateau pulled away from the stable mass of the continent and began to spin.

Whatever its origins, the Rio Grande Rift produced an arbitrary cut in the continent's interior, one that would help define the edges of the Colorado Plateau even further. Rift zones cut through continents elsewhere in the world—in the Great Rift Valley of East Africa, for example, and the Middle East's Dead Sea. While the rift's black lavas are easy to see, the faults that define its edges are largely hidden from view. Its borders are defined by a pair of more or less parallel faults that reach down several miles beneath the earth's crust, conduits for molten lavas and basalts from the fluid mantle below. In the mountains bordering the rift, mineral-rich fluids shot through the rocks, altering them in places to create deposits of turquoise. The pueblo peoples who settled here after the abandonment of the Four Corners area to the north turned these colorful sky blue pieces of rock into jewelry and ornaments. Pieces of it have been found as far south as Mexico City amid the ruins of Tenochtitlán, the Aztec capital. Later turquoise from the mountains of New Mexico would appear among the crown jewels of Spain as well.

With the onset of volcanism and rifting, tension began to slowly pull the land here apart. Blocks of the earth's crust between these two deep faults dropped downward by thousands of feet. At first these down-dropped blocks appeared as a series of isolated basins, occupying a common trend but having no connection between them. Later eroding sediments and erupting lavas would fill these isolated basins almost to the brim, creating a continuous pathway to the south. Strip away these filling layers of rock and ash from the rift near Albuquerque and you will find not a shallow trough but a deep desert valley, five times as deep as the Grand Canyon, its floor more than twenty thousand feet below sea level.

The river, meanwhile, began to head south through this string of basins, meandering back and forth across the low surface of the rift, alternately eroding and building up layers of sand, gravel, and volcanic ash. Unlike most other rivers on the Colorado Plateau, the Rio Grande did not carve its own path through the land here but followed the already established route of the rift. Traveling between Albuquerque and Santa Fe and farther north toward Espanola, you can see the face of this unstable valley fill in the badlands of brightly colored clay and ash that border the highway. Soft and easily washed away, erosion has carved them into a maze of pinnacles and arroyos, a collection of pink, tan, and buff-colored rocks.

Farther east in the Gulf of California, the East Pacific Rise initially created a similar set of deep basins. Widening the gulf, however, deflected the course of the Colorado River, turning it to the southwest, and the emerging sea that had begun to appear between Baja California and the Mexican mainland. Today the rocks that were carved from the Kaibab Plateau in northwestern Arizona to create the Grand Canyon lie at the bottom of the Gulf of California.

Patterns in earthquakes suggest that New Mexico's Rio Grande Rift is still active. Hundreds of small earthquakes shake the state each year, most too small to be detected anywhere but on the sensitive tracings of a seismograph. Ninety-five percent of those earthquakes are centered in a seventy-mile stretch of the rift between Albuquerque and Socorro. Within it, some twenty-five miles north of Socorro, the surface of the rift is actually rising at the rate of an inch every five years. While no major quakes have struck the region in recent times, archaeological and geological evidence suggests that earthquakes may have destroyed a number of prehistoric pueblos in the Rio Grande area.

South of Socorro the Rio Grande Rift spreads out into a broad band of faulting more than a hundred miles wide. Seams of lava still mark the surface here, but the parallel faults that define much of its reach are all but indistinguishable from the Basin and Range terrain that sprawls across southern New Mexico and Arizona below the Mogollon Rim. The same Basin and Range terrain then curves northward to enclose the western edge of the Colorado Plateau as well.

As it heads southward toward Mexico, the steadily widening rift spreads outward to include the Tularosa Valley and the barren deserts of the Jornada del Muerto. The Tularosa Valley contains the Trinity test site, the spot where scientists from the Manhattan Project, much of whose work had been done farther north at Los Alamos in the shadow of Valles Caldera, exploded the first atomic bomb and turned the floor of the desert to glass. The Jornada del Muerto is "the Journey of Death," a waterless route from the border of Mexico to the south, favored by Spanish missionaries and colonists heading north because it offered safety from the deadly attacks of the Apache who roamed the more fertile reaches of the Rio Grande.

Beyond El Paso the Rio Grande River heads toward the Gulf of Mexico along the U.S.-Mexican border. The rift, however, heads southward into Mexico through the harsh deserts of Chihuahua that lie between the rich, green world of central Mexico and the cooler land to the north. Once the Spaniards were established in Mexico, however, rumors of gold and lost cities would eventually draw northward, with dreams of finding a world even richer and more exotic than the one they had found in the south.

F I F T E E N

Pinyon and Juniper

SOUTH OF ZUNI the forest seems to go on forever. As the narrow two-lane highway heads south toward the tiny towns of Pietown and Quemado, it passes over a series of low hills and mesas. From the crest of these low rises you can look out over a dark green haze of pinyon and juniper that stretches as far as the eye can see, broken only by widely scattered reaches of open range: sagebrush and bunchgrasses and the conical shapes of distant volcanoes and cinder cones. It was the week after Shalako, and in nearly an hour of driving I had not seen a single car or house. The world seems to be made up of nothing but trees. Last night the temperature had dropped well below freezing; here and there among the trees were patches of snow. By late afternoon, however, the temperature would climb above fifty degrees.

Pulling off to the side of the highway, I parked my truck on an apron of gravel and began walking. A hundred feet into the trees and the road disappeared completely from view. The forest was low, its top little more than twenty feet above my head—dark green pinyon pines with short, straight trunks and yellow-green junipers whose twisted trunks and branches were covered with peeling strands of bark. The trees themselves were solitary and well spaced. In places the ground seemed almost parklike, the trees solitary and well-spaced like those of a carefully planted orchard. Overhead their evergreen boughs spread outward in a broad green crown, like the leafy canopy of a stunted maple or oak. The spire-like shape of pines and firs in the mountains above is an adaptation to heavy snows: Their shape sheds snow like a steeply pitched roof on a mountain chalet. These evergreen pinyons and junipers, however, are not part of a mountain forest but tough desert trees, as adapted to life in this dry landscape as the shadscale and sagebrush in the desert valleys below. The evergreen needles of the pinyon and the scalelike needles of the juniper are not adaptations to cold but to heat.

In places the crowns of the trees are woven together, making a solid roof overhead. Elsewhere, where the trees are more widely spaced, you can see this forest's tight links to the desert in plants that cover the forest floor: candelabra and prickly pear cacti; thickets of rabbit bush, sagebrush, and four-wing saltbush; and tufts of grama grass and needle-and-thread grass and patches of black cryptobiotic soil. The waxy outer surfaces of the pinyon's and juniper's leaves or needles help reduce water loss like those of other desert plants. Other tough, dry shrubs are here as well: thickets of serviceberry, mountain mahogany, and cliff rose. While the pinyon and junipers themselves are constant and unchanging, the forest floor beneath them changes dramatically from place to place with changing soils and rainfall.

Pinyon-juniper woodlands are one of the most common forest types in the American West, covering an estimated seventy-five thousand square miles. Shifting varieties of these trees are found all the way from the Rocky Mountains in Colorado to the Sierras in California, and from Canada to Mexico. On the Colorado Plateau they dominate the landscape at elevations between five thousand and seven thousand feet—the same reach of land where the Anasazi farmers found the needed balance between temperature and rainfall for their fields of corn, beans, and squash. They are found not only here south of Zuni out toward the Mogollon Rim but covering the tops of mesas, plateaus, and foothills almost everywhere in the Southwest: on the flanks of the Sangre de Cristos outside of Taos and Santa Fe to the east, on the tops of mesas in Dinosaur National Monument to the north, and along the South Rim of the Grand

Inside the Grand Canyon from the
Tonto Plateau

Late afternoon snowstorm in Zion Canyon

Winter sunrise, Bryce Canyon National Park

Hanging garden and tree,
Glen Canyon National
Recreation Area

The Green River making its way through Split Mountain, Dinosaur National
Monument

San Rafael River cutting through the
San Rafael Swell, Utah

Clay hills and badlands, central Utah

The Waterpocket Fold, southern Utah

Clouds and grassland, northeastern Arizona

Hunting scene pictograph, Nine Mile Canyon, Utah

Pueblo ruins, Mesa Verde
National Park

Summer flowers on the Wasatch
Plateau. Note snowbanks and
mountain forests

Diatremes and volcanic necks, northeastern Arizona

The Henry Mountains from Island in the Sky area, Canyonlands National Park

Lava and pine trees, El Malpais National Monument

Pinyon pine and sandstone, San Juan River area, southeastern Utah

Fall storm from Mesa Arch, Canyonlands National Park

Visitors taking in the view at Delicate Arch, Arches National Park, Utah

Spires and pines, Bryce
Canyon National Park

Canyon to the west as well. The critical factor is not latitude or longitude but elevation and rainfall.

Although a variety of pinyons and junipers are found throughout the West, just two species dominate the Colorado Plateau—*Pinus edulis*, the Colorado pinyon pine, and *Juniperus osteoperma*, the Utah juniper. The first Europeans called them simply pines and cedars and sprinkled the landscape with names like Pine Grove, Pine Valley, Cedar City, and Cedar Breaks that reflected their wide reach. While the wood of the juniper trees was aromatic and burned with a sweet smell, they are not truly cedars but cypresses, a family of trees whose descendants are found almost worldwide. Those biological details, however, were of little importance to the natives or early settlers who found these widespread forests of pinyon and juniper an invaluable source of food and supplies. While the juniper's durable heartwood could be used to make nearly rot-proof poles and beams, the pinyon pine's rich crops of nuts were a prized source of food. Over the course of several centuries man and trees would slowly evolve together.

Like the corn that would shape the world of the Anasazi, the pinyon pines evolved farther south in Mexico. Fossil evidence suggests that the first pines appeared in northeast Asia some 180 million years ago, when dinosaurs were first beginning to roam the earth. By 105 million years ago those early pines had split into the two major types known today, hard and soft pines. The split, some researchers have speculated, may have coincided with their migration out of Asia and into both different and rapidly changing climates. Early on in their evolutionary history a wide variety of pines appeared, some for mountains and deserts, others for shorelines and even tropical forests. The first pines may have reached North America before the Atlantic began to open. Others would later migrate from Asia across the subcontinent of Beringia during low stands of sea level. While some varieties spread across northern Europe, Greenland, and Iceland, others seem to have headed southward into Mexico and South America along the margins of the vast intercontinental seas that covered much of the Great Plains and the Intermountain West some 80 million years ago.

In the mild climate and rugged terrain of Mexico pines underwent an evolutionary explosion as dozens of new species appeared—

including the first pinyon pines. Sixty million years ago during the Paleocene North America began to grow both warmer and drier, the start of an interval of increasing heat and drought that lasted for some 30 million years and well into Oligocene time. As the climate began to warm, the pinyon pines began to migrate northward out of Mexico along with other drought-tolerant shrubs and trees like mountain mahogany, cliff rose, and gambel oak, all part of distinctive Madero-Tertiary flora, as scientists have termed it, out of which much of the present flora and fauna of the Intermountain West and the deserts to the south would evolve.

By the time these early pinyon pines began moving northward, they were very different from their ancient ancestors that had appeared several hundred million years before, during the age of dinosaurs. No longer fast-growing trees geared to compete in a moist shady forest, they had evolved into a slow-growing, short-trunked tree that used water sparingly. And as they moved northward, they became even more adapted to the dry desert weather. While the pinyons that migrated out of Mexico had their needles arranged in bunches of five, those that developed in the Intermountain West had their needles clustered in bunches of two or separated. Like sage-brush, pinyons are able to rapidly evolve, not by doubling their chromosomes or some other elaborate genetic trick involving entire chromosomes, but by point mutations—changes to a specific strand or piece of DNA.

In North America low-elevation woodlands like the pinyon-juniper forests found today on the Colorado Plateau date back to the Miocene, some 5 to 24 million years ago, although fossils as old as Eocene (some 37 to 58 million years ago) of what appear to be pinyon pines have been found in the Florissant and Green River formations of Colorado. In spite of that early arrival in the north, both the pinyon pine and other shrubs and grasses here seem to have moved around considerably. Just as ancient seas washed back and forth across the surface of the plateau, so too has changing climate and rainfall caused forests and deserts to move back and forth across the region. Over the past 2 million years, for example, there have been at least sixteen separate and prolonged glacial intervals—periods of glaciation and cooling that lasted longer than a hundred thousand years. During the peak of the most recent Ice Age some fifteen to twenty thousand years ago, the pinyon-juniper forests that now carpet the mid-reaches of the Colorado Plateau were forced down into lower elevations that are occupied today by desert grasslands and scrublands. In turn, much of what is now pinyon-juniper forest was occupied by subalpine conifers. During the peak of the Ice Ages, pinyon pines were found as far south as Yuma, Arizona, at elevations as low as eighteen hundred feet, an area today that is part of the Mojave Desert.

After the close of the Ice Ages the first pinyon pines began migrating back into the Colorado Plateau. The first returning pinyons seem to have reached the eastern edge of the Grand Canyon by some ten thousand years ago, and the slickrock deserts of south-eastern Utah some seventy-two-hundred to thirty-four-hundred years ago. While these dryland forests were migrating back into the area from the south, early man was filtering into the area from the north.

In contrast to the deserts below, the pinyon-juniper forest seems rich and green. Its alternating patterns of sun and shade are a welcome relief from both the dark mountain forests above and the open deserts below. Walking here you can see animals from either world: mule deer seeking shade from the midday heat, whiptail lizards and desert wood rats that dart across the ground. Up in the trees chickadees and blue-gray nuthatches flit from branch to branch like butterflies. In the dead skeletonlike branches of solitary juniper, ravens survey the desert like sentinels, their eyes watchful and alert. Such sightings, however, are few and far between—the sum experience of several hours or even days of watching. You can walk for hours through the forest here without seeing a single sign of life. The air is rich and per-fumed, filled with the resinous smell of the pines and incenselike scent of the junipers, but it carries only scattered sounds—the tap-ping of a distant woodpecker, the soft sigh of the wind blowing through the trees. The forest floor is littered with pinecones, while the sweeping, broomlike branches of the junipers are laden with thousands of hard silver-blue berries, but there seems to be no one at all to harvest this rich supply of food. Dead pinyons lay where they fell, broken into lengths by decay, while the trunks of junipers stand out like bones, bleached to a silvery white by the wind and the sun; persistent and upright long after the last hint of green has gone.

Without warning, however, the forest can suddenly come to life as flocks of pinyon jays wheel through the trees in groups of anywhere from several dozen to several hundred, stopping to feed on their hidden caches of food. Life here is not spread out evenly across the land but clustered and sporadic, as prone to boom and bust cycles as a mining town or an oil patch—a few weeks of profligate wealth followed by months of poverty and drought. The cyclical nature of life here has as much to do with the life cycle of the pinyon pine as the limits of the desert. While the varied cacti, grasses, and shrubs offer a variety of fruits, seeds, and leaves throughout the year, the key to life here is the pinyon pine and its rich supply of nuts.

Crops of pinyon nuts are available only every three years, longer if problems arise due to drought or insects. Their timing, however, typically varies from place to place. Thus, while any particular stretch of forest bears nuts no more than once every three years, there is always a crop available somewhere. For birds like the pinyon jay the trick is to find plateaus and canyons where the trees are ripe and ready for harvest. When the pine nuts reach maturity in August, the forest here is suddenly alive with animals: squirrels, chipmunks, wood rats, nuthatches, and jays—all of them combing the forest floor and the trees for fresh new cones laden with pinyon nuts.

When the shoots or buds of new cones appear, they are nothing more than tiny white dots near the ends of branches and twigs. They look almost like tiny needles and lie dormant until the following year. By the start of their second spring the year-old buds begin to grow, their color changing from white to yellowish green or reddish purple. Growing both longer and thicker, they finally emerge as tiny conelets, little more than a quarter of an inch in diameter, with the texture of a pincushion. Later that spring the scales of the conelets open up and the eggs or ovules inside are fertilized with pollen blown by the wind, starting the process that will allow them to grow into seeds or nuts. The conelet then closes up again and grows into a prickly brown sphere to wait out the winter. The following spring the cone, now in its third year, begins to grow—bright green at first and laced with sap. As they mature they grow steadily in size, turning brown and tan and opening their scales to expose the seeds inside—eight or nine per cone, with the same size and shape as the pit of an olive or date.

Of all the animals that feed on the nuts of the pinyon pine, none is more closely tied to it than the pinyon jay, a blue-gray bird the size of a small crow that lives all but exclusively in the pinyon-juniper forest. The relationship between the pinyon jay and the pinyon pine is one of the classic tales of natural history—a story of inter-weaving dependence and design. While the pinyon jay depends on the nuts of the pinyon pine to survive, the pine in turn depends on the jay to bury its seeds in the ground to guarantee its own survival.

In late August as the nuts begin to ripen, pinyon jays start to flock up, gathering together into groups of several hundred, flying from place to place as they begin the harvest, breaking open the still-green cones to collect the nuts hidden inside. The seeds are carefully checked as they are harvested—weighed in the bill (inedible seeds tend to be unusually light in both weight and color) and checked for softness. Some are eaten immediately, but others are carried away in the bird's esophagus, as many as twenty at a time, and buried in the ground—a cache of food to be used when times are lean. A single flock can bury literally hundreds of thousands of seeds, caching them typically in open areas on the south sides of trees, where snow seldom accumulates and the ground is quickly warmed in the spring.

In the desert moist ground for seeds is critical for survival, and the pine nuts that remain unburied seldom germinate. While the birds rely heavily on the nuts for food, they often bury two to three times as many seeds or nuts as they need—leaving hundreds behind to germinate and grow. While the large nuts are a rich supply of food for the jays and other animals, for those that remain planted in the ground their size also provides a rich storehouse of food for the emerging seedling, enabling it to grow long roots quickly in search of water while the seedling itself is still taking shape above-ground.

The cyclical nature of the nut crops, in turn, allows the trees to periodically produce bumper crops, flooding an area with so many seeds that the birds are unable to eat them all and end up caching far more nuts than they will use in the course of a year—thus ensuring the tree's long-term survival. The birds, in turn, are so closely tied to the cycles of the pinyon pine that they alter their behavior in bad years—either by breeding late when the crop is finally ready or in some cases not breeding at all. Some research has suggested that the bright green, glistening cones act as an aphrodisiac for the birds, triggering breeding. When cone crops fail, the birds may end up traveling hundreds of miles from their home territory in search of food. Flocking, in turn, is critical for their survival because it enables the birds to find food supplies that would be nearly impossible for a single bird to find on its own. At the same time, flocks offer protection from predators. Feeding groups are surrounded by sentinels posted on treetops that scan the sky for signs of hawks and owls, with as many as thirty to forty birds ganging up on a would-be predator that refuses to leave the area. The pinyon jays also have no feathers covering their nares, or external nostrils, an adaptation that possibly developed to prevent them from being clogged by the pines' sticky sap. Pinyon jays are part of the genus *Gymnorhina*, a Greek word meaning "naked nose." Apart from this interdependent relationship, the jays may have played a role in the evolution of the pines by preferentially selecting and planting seeds.

Man has a long-standing relationship with the pinyon pine as well. Prehistoric peoples in the Southwest seem to have made use of the pinyon pine almost as soon as they arrived. Deposits of seed husks and charcoal from pinyon pines some six thousand years old have been found on the floors of caves and rock-shelters in association with other human artifacts that date all the way back to Archaic times. Traditional stories among the region's native peoples emphasize both the pinyon's value as a food source and its antiquity. Navajo tales say that pinyon nuts were the original food of the gods. Among the Tewa of the Santa Clara pueblo near Santa Fe, pinyons are believed to be the oldest of all trees, the provider of *Tó*—the original food of people in the old days. It is not surprising that pinyon-juniper forests surround most of the abandoned Anasazi pueblos in the Four Corners region. Scattered archaeological sites in the form of small camps suggest that the Anasazi not only lived in the pinyon-juniper forests but traveled around them as well to collect and eat pinyon nuts.

Pinyon nuts are richly flavored and high in both fat and protein, their food value comparing favorably with that of walnuts. Traditionally, they were eaten raw, roasted, and boiled, and even ground into flour. Out along the western edges of the Colorado Plateau and out in the Great Basin the Paiute and Shoshone peoples stored them in clay jars, or *ollas*, and stone-lined pits. Later in the 1700s Spanish colonists and settlers collected the nuts as well, shipping them southward to Mexico along with brandy, wool, and tanned hides in exchange for needed supplies. Traveling across the Navajo reservation in late August and early September, you can still see groups of people out gathering pine nuts, their cars parked alongside the road as they wander through the forest with blankets and sticks, spreading the blankets beneath the trees to catch the cones they shake from their branches. Trading posts list prices for nuts, while other buyers park near the junction of highways and back roads, their windshields posted with signs: "We Buy Pinyon Nuts." In 1936 more than a million pounds of pinyon nuts from the reservation were sold to eastern markets.

While heavy grazing has reshaped the desert grasslands below them, it has also reshaped the pinyon-juniper forest. In its natural state the forest floor is rich and diverse, the trees as spread out and open as the groves of an orchard. Over much of the region heavy grazing has stripped the forest of much of its understory (while the trees provide shade for cattle and sheep, its understory of grass and shrubs offers food) and caused it to grow together in an almost impenetrable thicket that in places is floored by nothing more than bare dirt and snakeweed. Although still yielding rich crops of nuts, the forests today are far less diverse than they were in the past and cover a far greater area as well. Chainings, burnings, and seedings have been attempted to change this balance, but far too many have been geared only to cattle—leveling the forest and then sowing the bare ground with nonnative grasses like crested wheatgrass. While federal agencies like the Forest Service and the Bureau of Land Management attempt to restore a much-needed sense of balance to the landscape here, the mistakes of the past are hard to overcome and new attempts at managing the landscape are often viewed with understandable suspicion by both Native Americans and environmentalists.

In the beginning it was not just the nuts of the pinyon pine that were viewed as a rich source of food but the entire forest itself: the thickets of yucca, serviceberry, and Indian ricegrass found among the trees were a source of not just food but needed supplies as well. Pueblo peoples used the durable wood of the juniper for beams and poles in their pueblos and the sticky pitch of the pinyon pine to waterproof baskets and repair broken pots. Navajo, in turn, used the wood of the pinyon pine for making handles for tools and saddletrees, while its gum was used to coat flat stones to make a nonstick griddle; even rotted wood was put into use: ground into talcum powder for babies. Uses of the juniper tree were no less versatile. Its soft, smooth wood was easily worked and carved into utensils and tools; its shredded bark was smoked in cane cigarettes in place of tobacco or woven into rope and sandals; eaten as food among the Navajo, its berries were dried to make ghost beads that were strung into bracelets to be worn by children to keep them from having bad dreams. Among both the pueblo peoples and the Navajo there were a host of medicinal and ceremonial uses as well: salves and emetics, charcoal for sand paintings, wood for ceremonial fires and offerings.

Wherever forests of pinyon and juniper covered the ground, the ancient people could find everything they needed to survive. Today the signs of early man abound wherever these rich forests are found, not just those of the Pueblo, Navajo, and Anasazi, but also the Archaic and Paleolithic peoples who preceded them. In the rugged and rocky deserts of the Colorado Plateau the pinyon pine is a magical tree of life.

S I X T E E N

Spaniards and Navajos

SPANIARDS

THE NIGHT SKY above Cordova is black and clear and filled with thousands of stars. It is early April, and although the orchards and fields outside of town are already tinged with green, the night air is cold by morning the ground will be covered with frost. The smell of damp, rich earth mingles with the incenselike smell of burning juniper. A small Hispanic village, Cordova lies just off the "high road" between Espanola and Taos that winds through the foothills of the Sangre de Cristos, perched on the side of a steep ravine. Its tight cluster of tin-roofed adobe houses are black and quiet, lit only by scattered, solitary lights in kitchens and living rooms.

Listening carefully in the windless night, you can hear the musical sound of running water in the irrigation ditches that run through the fields outside of town and in the stream in the ravine below. At the center of a warren of narrow streets, the village church sits on the edge of a small plaza formed by the junction of several streets. Families and small groups of teenagers cluster together around the walls of the churchyard and lean up against the sides of neighboring houses. Watching and waiting they talk and laugh in a mixture of Spanish and English. Suddenly the talking stops. "They're coming," someone says in Spanish.

Rounding a corner a few blocks from the church, the procession appears, their path lit by lanterns and flashlights. A collection of old men and young boys, they walk though the winding dirt and cobbled streets of the village praying and chanting. At the center of the group more than a half-dozen penitents walk barefoot in loose-fitting white trousers, their faces hidden by darkness and black cloth, some being struck with whips made out of what appears to be woven yucca, others in the midst of some private penance of their own. At their center, a stooped and barefoot figure stumbles through the streets, carrying a wooden cross so heavy he seems barely able to drag it across the ground. Unlocking the church doors, they move inside to pray and then vanish again into the night. After they leave, the crowd outside passes into the church, filling its pews and benches to capacity. Santos, the hand-carved wooden figures of saints, adorn the small and simple altar of the church. After an hour's wait in the cold spring night outside, the church seems warm and bright.

Over the next several hours the penitents come and go, stopping inside the church to pray with the congregation by candlelight and then heading out to pray at shrines on the outskirts of the village, or so I am told by a man sitting next to me. No priests or friars direct the ceremonies here. The processions and prayers are all the elaborate rituals and traditions of the *Hermanos Penitentes*, a lay brotherhood of the Catholic Church that began in the small Spanish villages of northern New Mexico in the late 1700s.

It was late in the night of Good Friday, the day that for Christians marks Christ's crucifixion and death on the cross. Among the traditional Spanish villages of northern New Mexico the focal point of the Holy Week is not Christ's Resurrection from the dead on Easter morning but his suffering and death on the cross. From early morning until late afternoon the narrow highway outside of town had been clogged with traffic—a steady stream of both cars and people—heading for masses and processions in churches and villages up in the mountains to celebrate the Holy Day. A few miles downhill at the village of Chimayó the roads had been filled with thousands of pilgrims heading for El Santuario, a tiny pueblo church surrounded by rolling desert hills just outside of town. They walked singly or in groups of several dozen: a collection of young and old, some bearing banners from churches and parishes in Santa Fe and Albuquerque; some barefoot, others with canes and walking sticks. A few had been on the road for several days, walking from as far away as Belen a hundred miles to the south.

The church was built in the early 1800s by Don Bernardo Abeyeta. Ill and near death, Abeyeta was summoned to the site in a dream and told to dig in the ground. When he touched the soft earth he was miraculously cured. Construction of the small church took three years and was completed in 1816. In a small room off to the side of the altar, the builders left a small hole in the floor of the church, so that others could touch the sacred ground as well. Today pilgrims by the thousands come to the small chapel in search of blessings and cures. Crutches, canes, and letters of thanks from those who have been cured line the walls of a narrow adjoining room, simple testimonials to the power of faith. That morning in Chimayó a line of more than one hundred had extended out the church doors and through the gates of the adobe walls that surround the churchyard. Inside they paused at the altar to genuflect and pray, and then gathered up small handfuls and scoops of the sacred earth, putting them into glass jars and plastic bags to take with them.

As I headed toward Taos and a late afternoon appointment, each village alongside the highway seemed to be busy with traditions and ceremonies of its own. At Truchas segregated processions of men and women marched through the village streets carrying statues of Christ and the Virgin Mary in a reenactment of *El Encuentro*, Mary's encounter with Christ as he carried his cross through the streets of Jerusalem on his way to his Crucifixion at Calvary. That afternoon at Talpa, closer to Taos, men and boys in dark suits and women and girls in bright dresses knelt and prayed outside the church to observe the *Via Cruce*, the Way of the Cross, a collection of prayers and readings that recall Christ's death on the cross in mid-afternoon.

This nighttime ceremony at Cordova is known as *Las Tinieblas*, the darkness, and marks the close of the Holy Week for the penitentes, commemorating not only Christ's death but his departure from the earth and descent into death. When the penitentes enter for the last time, the candles are slowly put out, leaving the church in total darkness. The darkness is then suddenly broken by an explosion of sound—a cacophony of whistles, rattling maracas, and shouts symbolizing the chaos that followed Christ's death. The candles are then slowly relit. Alternating periods of light and dark punctuated by sound and chaos continue for more than an hour. In between, the penitentes recite the names of the village's dead from memory, a roll call of those who have passed on.

When the ceremony comes to a close, it is well past midnight. The penitentes are the first to leave, filing quietly out of the church. They wait outside in the churchyard to greet members of the congregation, receiving thanks for their remembrance and prayers. Talking and laughing, the small crowd spreads out through the streets, back to their homes and sleep.

Spanish culture and language have left a deep and almost indelible mark on the peoples of northern New Mexico. For Spain, however, the land here was never much more than frontier, a buffer zone between the rich silver mines of Mexico, which were the basis of Spanish wealth and power in both the Americas and Europe, and the unknown and uninhabited lands to the north. Spanish control here in the Southwest lasted for nearly three centuries, reshaping the lives and destinies of its native peoples. The first arrivals were priests and soldiers, but in time settlers and colonists would come as well. Separated from the outside world by high mountains, deep canyons and deserts and several hundred miles from the nearest source of supplies in Mexico, they were forced to depend almost entirely on the land around them to survive. By the early 1800s a network of small, independent Spanish villages had appeared along the banks of the Rio Grande and its tributaries in northern New Mexico and southern Colorado. While their cultural roots and religion came from Spain, in time their lives became as closely tied to the land as those of the native peoples around them.

When Spain's empire in the New World began to unravel, the steadily shrinking supply of friars and priests forced Spanish colonists and settlers here to try and fulfill their spiritual needs as well. Groups of village men began to band together in religious brotherhoods known officially as *La Cotroadía de Nuestra Pater Jesus Nazareno*, or the Hermanos Penitentes, to preserve their religious values and customs. Organized into village groups, or *moradas*, members tried to imitate Christ's life not only through penance and suffering but in acts of charity and aid to their fellow villagers and brothers. After Mexican independence in 1821 the Southwest would sink even further into obscurity with regard to Spanish culture. When the area was finally seized from Mexico by the United States after the conclusion of the Mexican-American War in 1848, the region's Hispanic peoples would come to feel as threatened and dispossessed as its native peoples had felt after their own arrival several centuries before. Over the next several decades the elaborate rituals of the penitentes would provide a link with rapidly disappearing past.

In both Spain and the New World, 1492 was a landmark year. While Christopher Columbus's voyage to America marked the end of ten thousand years of separation between the Old World and the New and the beginning of European domination in the Americas, in Spain 1492 marked the end of more than four hundred years of domination and religious persecution at the hands of the Islamic Moors as the last Moorish forces were driven back to Africa. Finally free and independent, Spain was set to expand its reach around the world, moving with the same ruthlessness and determination that had once driven its own enemies.

The search for wealth and power was a critical force behind Spain's drive to conquer the New World. "We have," Hernán Cortés would explain to the Aztec emperor Montezuma, "a strange disease of the heart for which gold is the only cure." It was an article of faith that things that were rare in the Old World were plentiful in the New. Landing on the Mexican coast near Veracruz, Cortés had his men burn their boats, leaving the wary among them with no possible means of turning back. When they reached the Aztec capital of Tenochtitlán, located where Mexico City now stands, they found a world beyond their wildest dreams—a sprawling city of more than a quarter of a million, laced with canals and floating gardens of flowers. Far larger than any European city of its day, its center was marked by giant pyramids whose size and scale were comparable to those of ancient Egypt, while palaces and vast aviaries of exotic birds lay nearby. Its treasuries were laden with stores of gold and precious stones. Less than thirty years after Columbus had reached America, Cortés would topple the ancient empire of the Aztecs and the world of New Spain would emerge, its churches and palaces built directly on top of the pyramids and temples it had replaced.

Cortés's success sent a wave of other would-be conquistadors in motion, claiming the lands and people they encountered (and sometimes conquered) for God and the Spanish king. Few, however, would end in success. In 1528 a would-be conquistador by the name of Pánfilo de Navárez led an expedition of four hundred to the Florida coast. Landing near Tampa Bay they pushed inland, only to be decimated by hostile natives and frustrated by the thick and almost impassable swamps of the interior. They retreated back to the beach but found that their boats were gone. Building their own makeshift rafts, they set sail across the Gulf of Mexico hoping to reach Mexico, but ended up shipwrecked on the Texas coast near Galveston. Battles with the natives, disease, and starvation quickly took their toll. Only four of the party's original four hundred survived, but it would take them more than eight years to make it back to Mexico City. No longer conquerors but captives, they were kept as slaves by natives along the Texas coast. After several years in slavery they would finally make their escape, working their way across Texas, Arizona, and New Mexico before finally turning southward into Mexico. Two of the four would become well known in history—a Spanish nobleman by the name of Alvar Núñez Cabeza de Vaca and a black Moor by the name of Esteban de Dorantes, a slave who had been brought to the New World from Spain. Taking on the trappings of shamans and healers, they slowly worked their way westward and southward, passing from tribe to tribe, learning their customs and culture.

Their return to Mexico City created a sensation, and their stories of deserts, mountains, and vast plains dotted with herds of cow-like animals that would later become known as buffalo intrigued local officials and bureaucrats. But even more enticing were the rumors of towns to the north, cities of stone houses whose residents were said to possess riches of turquoise and gold. While the route taken by Cabeza de Vaca and the others had not brought them into contact with those northern cities or pueblos, running as it did across the southern edge of New Mexico and Arizona, their reports of stone cities gleaned from natives and locals led many to believe that there was another hidden world of wealth waiting to be discovered in the unknown lands to the north.

The pueblos, many Spaniards assumed, were the Seven Cities of Cíbola, part of a mystical kingdom founded by seven bishops who had fled Spain in the 1200s to escape the advance of the Moors. The story had begun more than four hundred years before with the Moorish capture of Mérida. According to the legend, the bishops and their followers had escaped the town's fall and fled by ship across the western ocean, landing on a chain of blessed isles after a journey of several weeks, where they built a utopian community of unimaginable wealth and sophistication.

The possibility so intrigued the viceroy of New Spain, Don Antonio Mendoza, that he asked Cabeza de Vaca (whose name means "head of a cow" in Spanish) and his fellow survivors to lead a small reconnaissance mission northward in search of these unknown towns to see if they were in fact the fabled cities of Cíbola. All save Esteban, the black Moor, respectfully declined. In their place Mendoza appointed an ambitious young cleric, Fray Marcos de Niza, to lead the party and sent Esteban along as his as guide.

In 1538 they started out on foot from Culiacán in the coastal Mexican state of Sinaloa. Near the Rio Fuerte, Marcos and Esteban parted company. While Esteban pushed on ahead, Marcos stayed behind to make notes on the nearby Pacific Coast. Left to himself, Esteban traveled in style. Passing from village to village on his way north, he dressed as a medicine man, adorned himself with jewelry and feathers, and began to acquire both a following of warriors and a harem of young women. Accepting gifts of turquoise and distributing blessings as he headed towards Cíbola, Esteban moved northward like a Pied Piper, drawing a string of followers behind him. Marcos, the nervous young cleric, followed in his wake, always a few days behind.

In spite of his talent for both attracting and seducing his followers, Esteban could neither read nor write. In place of written notes, he and Marcos had devised a code to report on his findings. If he found a moderately sized city, he was supposed to send back a cross that was a span across (a span being roughly nine inches), two spans if he found something more interesting. He was to send back something larger only if he found something comparable to the Aztec capital of Tenochtitlán. As Esteban approached the pueblos of Zuni he sent back a cross as large as a man. The first contact between the Old World and the New in the Southwest would not be with a white European but a black African who went off in search of Cíbola, but found the descendants of the Anasazi instead. For Esteban, how-ever, that first contact would be his last.

Receiving Esteban's gigantic cross, Marcos was ecstatic and hur-ried forward to see things for himself, only to be greeted by frantic members of Esteban's party fleeing Zuni in terror. Esteban, they said, had been captured and killed. The Zuni had bought none of his posing as a healer or leader and spurned his demands for women and jewels. To prove that he was human they had cut his body into strips and then distributed the pieces among the village elders. After receiving the news, Marcos claimed to have gone on ahead and studied the pueblo from a distance. Whatever the facts of the matter, by the time he returned to Mexico City, his observations bore little relationship to the truth—other than reporting that Esteban was dead. They had, he said, found a city larger than Mexico City itself, and furthermore the trip was easy. Encouraged by the news, the viceroy would appoint a young Spanish nobleman by the name of Don Francisco Vásquez de Coronado to lead an *entrada* to the north and claim these rich new cities for Spain.

Coronado headed north in 1540 with a group of more than a thousand men—some three hundred Spanish soldiers and several hundred Indian auxiliaries and servants. Along with them were an even greater number of horses and cattle, perhaps several thousand head in all. They had dreams of a great conquest, a victory even greater than Cortés's conquest of Mexico. The trip, however, proved almost disastrous from the start. They had a torturous trip through the mountains of northern Mexico and then trudged through the hot deserts of southern Arizona before finally climbing up into the cool forests of the Mogollon Rim and heading east toward Zuni. The long trip had broken almost all of their spirits, and what they saw as they peered down into the pueblos of Zuni broke the rest. Before them lay not an imperial city of pyramids and canals but what Pedro de Casteñeda, one their number, would later describe as "a little crowded village, looking as if it had been crumpled up all together." Instead of the Seven Cities of Cíbola with streets paved with gold, Coronado and his men found a collection of six small pueblos built out of dried mud. After the wealth and splendor of Mexico, their disappointment must have been acute.

Coronado's men reached the Zuni pueblo of Hawikuh and through an interpreter asked for both peace and submission. As they approached, the Zuni sprinkled a line of sacred cornmeal on the ground and warned them not to pass. Spanish pleas for peace were met by a shower of arrows. Exhausted and sick from their long trek, Coronado and his men had hoped to avoid a battle but they had no thought of retreat. Charging the Zuni warriors on horseback they sent them fleeing helter-skelter back to the pueblo. Reaching the pueblo walls they dismounted and continued the attack on foot, only to be pelted with arrows and stones. Coronado himself, arrayed in fine armor, quickly became a target and was knocked unconscious by a hail of stones thrown from the walls above. Spanish harquebuses, primitive muskets, finally saved the day for the outnumbered Spaniards. With Coronado unconscious, one of his captains managed to lead a successful raid on the entrance and overran the pueblo. The Zuni pleaded for mercy. Told they were welcome to stay as long as they remained peaceful, they opted instead to abandon their pueblo and disappeared into the surrounding mesas and hills. Inside the Spaniards found not mounds of gold and turquoise but something far more valuable in their exhausted state—food, several thousand bushels of corn and beans as well as dozens or perhaps even hundreds of turkeys. After settling in and catching their breath, Coronado sent messengers around to the other nearby pueblos—now all abandoned—and let it be known that they were eager to meet with their leaders and learn more about their country. Eager for peace, the Zuni pledged allegiance to the Spanish king and expressed, or so the Spaniards thought, a desire to become Christians.

Over the next few months these episodes of brief and bloody conflict and sudden reconciliation would be repeated to the east and west as Coronado sent expeditions out to explore the unknown land around them. To the east they would travel as far as present-day Albuquerque and Santa Fe, making contact with more than eighty pueblos along the banks of the Rio Grande and the Pecos rivers before making a brief foray out into the Great Plains—all in the space of a few months. To the west they would travel to Hopi, which they called Tusayan, and find guides who would take them to the South Rim of the Grand Canyon near Desert View. Hoping that the river might eventually lead them to the Gulf of California, where supplies were being sent by ship from Mexico to aid the expedition, they tried unsuccessfully to reach the canyon floor. The supplies never reached them, although the ship's crews would manage to follow the Colorado River all the way to the mouth of the Gila. Miraculously enough Coronado's men would drop down off the high plateaus of northern Arizona and follow the Colorado River southward and locate not only the mouth of the Gila but the letters and messages left for them by their would-be supply ship.

With winter approaching, Coronado decided to move his forces out of Zuni and over to the Rio Grande, where the land was richer and the pueblos were larger. With the first winter snows beginning to arrive, they ordered the inhabitants of Alcanfor near San José to abandon their pueblo and then dug in to spend the winter, warm and secure in the natives' pueblo, living off their carefully stored supplies of food. The transition from explorer to invader was now complete. In a matter of months the Spaniards and the natives would be at war: raids on horses and herds of cattle by the natives followed by bloody sackings of pueblos by Spanish forces. Pueblos were surrounded, their inhabitants starved into submission and then, on at least one occasion, burned alive after they left unarmed under assurances of amnesty. By spring more than twelve pueblos had been destroyed. Others nearby were abandoned as their inhabitants fled into the desert.

In spite of all evidence to the contrary, neither Coronado nor his men had given up their dreams of finding lost cities and hidden stores of wealth. Unable to defeat the Spaniards in battle, the natives soon learned to make use of this weakness to drive them from their midst, regaling them with stories of fabulous cities and kingdoms out in the Great Plains, a mythical world that the Spaniards came to call Quivira. The following spring the pueblos would give them a guide, a Plains Indian they had captured several years before and held as a slave, giving him his freedom in exchange for leading the Spaniards away and, if all went well, to their death—far from food and water in the broad, flat land to the east.

They left in late spring, wandering across the vast, open reach of the Great Plains like sailors lost at sea. There were no trees or hills to mark the horizon; no recognizable points of reference, only an endless ocean of grass. Members of the group sent off to hunt or scout lost all sense of direction and never returned. They made contact with wandering bands of Apache (whom they called Querechos) following herds of buffalo, but found no signs of Cíbola or Quivira. From early May until late July they wandered in circles out on the Great Plains before realizing they had been deceived. El Turco, their guide, was garroted and buried. Coronado and his men then returned to the Rio Grande and spent another winter among the pueblos. In the spring of 1542 they informed the natives they were free and headed south in disgrace. News of his failure and treatment of the Indians would not go down well in Mexico City. Coronado would be tried for both mismanagement and cruelty to the Indians and die with both his health and fortunes irretrievably ruined. Spanish forces would not return to the Southwest for more than fifty years.

In spite of Coronado's spectacular failure in the Southwest, rumors of lost cities and hidden treasures of gold in the north would prove almost indestructible. But while the promise of easy riches lived on in the imagination, travel to this still unknown region was carefully regulated by Spanish colonial authorities, possible only with official approval. With their hands full in Mexico, colonial authorities had little interest in pushing the frontier notrthward; nor did they wish others to repeat Coronado's brutal battles with the Indians. Over the next few decades a handful of self-appointed conquistadors made the long trek north into the Southwest, only to be pursued by Spanish soldiers and brought back to Mexico by force or wander back under their own power exhausted and disillusioned. Others simply vanished, heading out into the Great Plains, and were never seen or heard from again.

After several appeals, the petitions of Don Juan Oñate to establish a colony in the north were finally approved in 1595. Oñate was governor of the colonial prince of Zacatecas, and his wife was a direct descendant of both Hernán Cortés and the Aztec emperor Montezuma. Oñate was so obsessed with the Southwest that he had even offered to finance the costs of the expedition himself, but approval came slowly. Spanish bureaucracy was almost as fearsome as the Spanish Inquisition, and neither wealth nor position nor official permission would entirely free him from its clutches. When they finally made their departure after nearly three years of delays, they headed north only to be held up for weeks in the colonial frontier town of Culiacán while authorities took an inventory of their stores and supplies—recording everything Oñate and his colonists carried with them down to the last nail.

On May 1, 1598, they forded the Rio Grande near El Paso and began following the river northward into New Mexico. The caravan was more than four miles long, a column of nearly a hundred heavily loaded wagons followed by more than seven thousand head of cattle, sheep, and horses with a collection of 130 families and more than two hundred single men scattered among them. There was nothing secretive about their journey, and it did not go unnoticed. As they headed north they found the pueblos alongside the river had been abandoned before them. Although more than a half century had passed since Coronado's departure, news of the Spaniards' return had caused the people to flee. Finally Oñate sent a smaller advance party ahead to assure the natives that they came in peace and to distribute trinkets to calm them. They settled near present-day Espanola, a few miles northwest of what would later become Santa Fe. At the junction of the Rio Grande and the Rio Chama they "converted" the Tewa of Yunque to Catholicism and then threw them out of their pueblo, moving in for the winter and claiming the food and supplies they found inside for their own, renaming it San Juan de los Caballeros—the Knights of St. John.

Oñate quickly divided New Mexico into mission districts and then set out the following spring to find the wealth that had eluded Coronado, leaving a bloody trail of conflict behind him. In the space of a few months his troops would kill more than a thousand. When residents of Acoma killed one of Oñate's favorite nephews, he all but destroyed the pueblo. When the siege was over, more than six hundred had died. The warriors who survived were sentenced to twenty years of slavery and had one foot cut off. Two Hopi who had been caught fighting alongside the Acoma had their right hands cut off and were sent back to their pueblos to teach the others a lesson. His brutality was numbing. A reign of terror had begun.

Oñate proved almost as brutal and incompetent at managing his own colonists as he did the pueblos. After an unsuccessful foray out into the Great Plains, Oñate and his soldiers returned to San Juan to find their newly created capital almost completely abandoned. In his absence most of the capital's clerics, colonists, and soldiers had decided to return to Mexico. The Southwest seemed poor and bare after the riches of Mexico. The winters were bitterly cold, while the ground was scarcely able to produce a good crop, much less silver or gold. While others were eventually brought up from Mexico, they too would soon begin to leave. Those who were caught by Oñate's soldiers were executed as deserters. In less than ten years he would be recalled to Mexico in disgrace and stripped of his titles for his mistreatment of both the natives and the colonists.

But while Oñate was replaced and the capital moved to Santa Fe, the die had been cast. For the next several decades a string of unsuccessful governors would rule New Mexico, each one seemingly more incompetent than the last. Those who came to rule the New World were not the flowers of Spanish chivalry but the weeds—brutal men who had been sent off to Mexico and the Americas by their families to escape punishment from the Inquisition or to make their fortunes. While governors and soldiers forced the natives to work in their fields, friars and priests tried to force them into the church, burning kachina masks and flogging their religious leaders for idolatry after urging them to kneel and kiss the cross.

By the 1630s church and civil authorities were at war among themselves—each one accusing the other of abusing the natives. In 1637 the governor, Luis de Rosas, set up a series of sweatshops in the Palace of the Governors at Santa Fe, promising the natives religious freedom in exchange for weaving blankets and tanning hides. When the church sent two friars to register a complaint, de Rosas beat them senseless with a stick until they were bathed in their own blood. Church officials retaliated by sending out men to pillage de Rosas's ranch. De Rosas then countered by sending out groups of natives to defile and rob churches.

Other problems soon followed. By 1640 disease was beginning to ravage the area as well. Throughout the Middle Ages epidemics and plagues had killed millions in Europe. In the 1600s smallpox, measles, whooping cough, and cholera would decimate the New

World; imported diseases for which the natives, unlike the Spaniards who unwittingly carried them, had no natural immunity. In the southwest the pueblos were decimated. The western pueblos of Zuni, Acoma, and Hopi lost an estimated 75 to 80 percent of their population. Others along the Rio Grande were abandoned as epidemics killed thousands. A three-year drought followed in 1665, and famine began stalking the pueblos along with disease. Villages and pueblos were abandoned as the Indians wandered in search of food. Travelers reported that roads and trails were littered with the bodies of these who had died of starvation. On their ranches and missions the Spanish were reduced to eating strips of parched cowhide. Apache, Navajo, and Ute, badly in need of food, stepped up their raids, killing and stealing and adding to the rapidly growing disaster. With memories of both past and present wrongs strong in their thoughts, religious leaders among the pueblos blamed their problems on the aliens among them. Their difficulties, they said, had all begun with the arrival of the Spaniards. Because of the Spanish and their alien god, their own gods and kachina spirits had refused to send either rain or blessings.

Faced with growing unrest, nervous priests and friars tried to tighten the screws. In 1678 they arrested forty-seven pueblo religious leaders. Three were hung. One committed suicide. The rest were whipped. The numbers, however, were not in their favor. While drought and disease had reduced the pueblo population from more than forty thousand at the time of contact to no more than fifteen thousand, the total non-native population of the colony numbered no more than thirty-five hundred. While more than a thousand clustered around the capital at Santa Fe, the rest were scattered across the region in isolated ranches, haciendas, and missions. All told they had an army of less than two hundred professional soldiers. The Spaniards, as historian John Upton Terrell remarked in his book *Pueblos, Gods and Spaniards*, had built a bomb. In 1680 it would explode.

By 1679 the pueblos were openly calling for revolt. Their anger had simmered for nearly a century, but old animosities and persistent divisions had always kept them from joining together. Out of San Juan Pueblo, however, a leader would emerge, the catalyst who would bind them all together. Known as Popé, he had been among those religious leaders arrested and whipped by the Spaniards in 1678. Claiming to be in contact with spirits from the underworld from a kiva at Taos Pueblo, Popé began planning the uprising, sending messengers and holding meetings with villages and tribes throughout the region and building up a network of alliances that reached all the way to Hopi more than four hundred miles away. Ute, Comanche, Navajo and Apache would join in the attack as well.

To coordinate the uprising he sent a knotted cord to each of the pueblos. Each day at nightfall they were to untie a knot. When the last knot was done the revolt would begin and they would drive the Spaniards from their midst. Secrecy was everything. Popé went so far as to kill his own son-in-law when he suspected him of treason. A few days before the attack was scheduled to begin, three leaders of the Southern Tewa went to the Spanish governor in Santa Fe and warned him of the coming attack, but their news came too late. Learning of the treachery, the pueblos launched their attack a day early. On August 9 at the Tesque a Spanish priest was cut to pieces as he prayed. From there the violence spread like a fast-moving fire. Over the next two days seventy Spaniards were killed at Taos, twenty at Picuris, and dozens of others at San Ildefonso, Nambe, Pojaque, Santa Clara, San Juan, Santo Domingo, Santa Ana, Sandia, Cochiti, San Felipe, and elsewhere. Stripped and mutilated bodies lined the roads and lay in the courtyards of haciendas. To the west at Acoma, Zuni, and Hopi, every church was destroyed. In less than three days more than four hundred men, women, and children were killed and twenty-one priests were martyred. Young Spanish women were taken alive and turned over to the pueblo leaders for their amusement.

At Santa Fe more than a thousand colonists were trapped inside the Palace of the Governors, where they had taken refuge. With a force of no more than 120 soldiers to defend the palace walls, they were surrounded by more than two thousand Pueblo warriors. Taken completely by surprise, most had only the ammunition they happened to be carrying in their pockets and pouches at the time of attack. After three days the irrigation ditch that carried water into the compound was shut off by the encircling Indians. For the next week they huddled in fear and watched as their horses and animals died of thirst while others went completely insane and had to be killed. After nine days the Spanish leaders decided it was better to die fighting than of hunger and thirst in the *casas reales*. The out-numbered soldiers attacked and broke through the Indian lines, killing more than three hundred and putting the rest to flight. On August 17 the surviving Spaniards headed south on foot. By the end of September they finally reached El Paso. Among their ranks were natives from pueblos to the south who had converted to Catholicism and now feared for their lives. Women and children of the group, it is said, looked like the walking dead. In Mexico both the defeat and the death toll were considered staggering.

In spite of the intense pressure to retake the colony, outside of a few desultory expeditions up the Rio Grande, Spanish forces would not return to New Mexico for twelve years. Spanish colonial leaders in Mexico City had other, more pressing concerns: a growing Indian insurrection within Mexico itself and increasing pressures from other European powers as well. By the late 1600s French forces were laying the groundwork for their own colonial empire along the Mississippi River while English forces in the Carolinas were inciting the Indians there to attack the Spanish in Florida.

After they had driven out the Spaniards, Popé ordered his fol-lowers to go to the river and bathe, scrubbing themselves with yucca soap to wash away the contamination of baptism. People were urged to rid themselves of every sign of Spanish presence: orchards, vineyards, cattle, horses, and sheep. Churches were burned and defiled, missions destroyed. Symbols, signs, and artifacts, however, proved far easier to destroy than ideas—and not all were eager to abandon the new ways of life they had learned with such great pain from the Spaniards. The world had changed. There was no way of turning back.

In the end it was not a search for wealth that drove the Spaniards back to New Mexico but a search for souls. Having baptized the pueblo peoples and others into the church, they felt obligated to return and save their souls from what they perceived as a fall from grace into infidelity and the pagan rituals of the past. In 1692 the newly appointed governor of New Mexico, Diego de Vargas Ponce de León, left El Paso with a force of just 160 men to reclaim the land to the north for Spain. Son of a prestigious family in Spain, de Vargas had left a wife and a secure life in the royal court to seek adventure in the New World. Successful in business and deeply religious, he had offered to finance the expedition himself. Of his group only thirty to forty were regular troops. The rest were Indian auxiliaries from Mexico and volunteers from the ranks of the colonists who had been driven from New Mexico twelve years before.

Heading north in a slow-moving column, they left the bulk of their wagon train near Socorro some hundred miles south of Santa Fe and began moving rapidly up the Rio Grande. Smoke signals sent from the tops of nearby mesas told them their movements were being carefully watched. When they reached the pueblos of Cochiti and Santo Domingo outside of Santa Fe, they found them completely abandoned. They rode into Santa Fe on the morning of September 13 just before sunrise and found that the former Palace of the Governors alongside the central plaza had been converted to a pueblo, its windows and doors sealed by bricks and adobe. From its roof a crowd taunted them with hoots and hollers. De Vargas demanded their surrender and told them they had an hour to think it over. While the Indians considered their options, his men cut off the compound's water supply and positioned their cannon for the attack.

When they decided to sue for peace, de Vargas put aside his armor and put on the silk of the royal court and went inside. Over the next two months the pueblos surrendered one by one. Taos, where the revolt had been planned, was captured without firing a shot. At Acoma, where Spanish priests had been tossed from the cliffs, more than a hundred children were baptized. While the mission church at Zuni had been completely destroyed, all the property and sacred objects of the missionaries had been saved. Led to a small room by the Zuni, the Spaniards found candles burning on a simple altar. The Zuni themselves offered no explanation. In November de Vargas assured the pueblos he would return soon with others to live among them again and started on the long journey back to Mexico. By December 10 he and his men were back in Socorro. Five days before Christmas they reached El Paso and crossed the Rio Grande into Mexico.

Politics and an ongoing war with the Apache in northern Mexico delayed de Vargas's return for nearly a full year. In early October orders from the king of Spain finally arrived, and de Vargas finally began heading north with a small army of some eight hundred colonists, soldiers, and priests. Along with them were more than two thousand horses and nine hundred cattle as well as herds of goats and sheep. On December 16 they reached Santa Fe. The padres entered the plaza singing but found several hundred Indians watching them silently from the roofs of the former palace. De Vargas asked for friendship and loyalty but received no response. Rather than drive the natives out that day they told them they would have time to leave. With winter settling in around them the Spaniards moved to a camp outside of town. They waited for eleven days. Twenty-one died of exposure and cold. On December 28 de Vargas went back to the plaza and again requested that the Indians leave. One of their leaders broke the silence, warning de Vargas to go back to his fraudulent god and Maria or stay and be killed, sending him off with screams and curses.

The next morning de Vargas and his soldiers kneeled in the snow to pray. Bearing the banner of the Virgin Mary, they stormed the pueblo walls, shouting that they were fighting for the Blessed Sacrament. The Indians fought back fiercely, but cannon fire and mines set by sappers finally breached the palace walls. With night falling around them, the Spanish forces withdrew to their camp outside of town and huddled around their campfire. When they went back to rejoin the battle the next morning they found that the Indians inside had made no effort to repair the walls. The pueblo surrendered, but de Vargas's patience and mercy had worn thin. He had seventy men executed and went to sleep that night inside the walls of the Palace of the Governors with statues of Jesus and La Conquistadora at his side.

The Pueblos submitted quickly to the return of Spanish rule. While oppression had united them in the past with both each other and their traditional enemies the Apache, Comanche, Navajo, and Ute, twelve years of freedom had caused their unstable alliances to unravel. While outside tribes stepped up their attacks and raids, the Pueblo peoples began to bicker among themselves. Drought had preceded the Pueblo Revolt and droughts would follow it as well. The old ways, it seemed, no longer held their power. Villages split up into pro- and anti-Spanish factions. Up in Taos, Popé, the leader of the Pueblo Revolt, settled in to rule like a colonial don, demanding tributes from neighboring pueblos and maintaining a harem of women he had pressed into service. Rather than the leader of a religion that emphasized community participation, he claimed to be an earthly emissary of the supernatural and that the details of things were revealed to him in private kiva séances. Like other dictators both before and after, he would die almost completely insane.

To outsiders the Southwest's pueblos seem almost identical, with similar styles of architecture, art, and agriculture that reach all the way from the Rio Grande to Hopi. In point of fact, however, the Pueblo peoples are made up of no less than six separate groups, each with its own distinctive language and history: Tewa, Tiwa, Keres, Jemez, Zuni, and Hopi. Not only are these different groups or tribes distinct, but the villages within them are often surprisingly independent as well—small city-states that were in turn divided by their own complex collection of clans and kiva societies. While the elaborate rituals of the kachina cult and their native religious held them together, like any close-knit isolated group there were inevitable conflicts and animosities bubbling away beneath the surface; disagreements between not only villages and clans but families and individuals.

Human memory is a tricky thing, capable of both suppressing the past and focusing upon it, compressing the wrongs and abuses of several centuries into an endless litany of suffering. Early histories of colonial times often ignored the terrible episodes of violence and persecution that marked the first contact between Europeans and Americans. Revisionist histories today focus almost entirely upon them, suggesting that whippings, burnings, and executions were a weekly, if not daily, occurrence—compacting the events of several centuries into an endless catalogue of abuse. Such interpretations are no less inaccurate than those that preceded them. The relationship between Spaniard and Indian was far more complex than that of oppressor and oppressed. While both could explode into violence, more often they simply lived together. However uncomfortable or unwanted, the common ground beneath their collective feet would slowly bring them together.

Decades before the arrival of Hernán Cortés and his small army of conquistadors, Aztec prophets had predicted the arrival of a lost brother, a light-skinned god from across the water, *Quetzalcoatl*, the god of civilization. His arrival, the prophets said, would come in the year *Ce Acatl* on the Aztec calendar—1519 A.D.—the year of Cortés's arrival. To the south Mayan theologians had made similar predictions of their own; so too had tribes in the Southwest. Among the Hopi he was known as Páhana, a long-lost white brother who had been separated from them when they emerged into the Fourth World and began the wanderings that would eventually bring them to their high, dry mesas in northern Arizona.

It would take Cortés more than a decade to conquer the Aztecs. His victory would not come solely with a force of brutal and well-armed Spanish soldiers as is commonly reported but with the help and aid of an army of several thousand Indians, drawn from the ranks of neighboring tribes and peoples who had chafed under the brutal domination of the Aztecs. (Coronado, Oñate, and de Vargas would likewise fill the ranks of their own forces in the Southwest with native peoples from Mexico.) While the Aztecs built a sophisticated and beautiful civilization with an artistic and architectural tradition that equaled anything in Europe, they were no less addicted to war and violence. Believing that human hearts were necessary to keep the sun in motion, they sacrificed captives from the tops of their temples on a daily basis, victims drawn from the ranks of the weaker tribes and peoples around them, from whom they also exacted crippling tributes of not only food and gold but forced labor as well. Cortés would bring the scattered enemies of the Aztecs together and lead them into war, not the source of the explosion that would rock ancient Mexico, but merely its trigger. Cortés would play his part as Quetzalcoatl only as it suited him. The opportunity for a golden age of peace and friendship that the old myths had implied would be lost for good—a tragedy that would be repeated over and over again in the Americas as the Old World came into contact with the New.

If European arrival in the New World had been postponed for a century or more, or Montezuma's forces had been able to drive the Spaniards from Mexico, the history of the New World would have been far different—but perhaps no less violent or painful. The simple traditional societies of the Southwest would have been easy pickings for the organized and warlike Aztecs. Instead of building churches for Spanish priests and carrying water for Spanish soldiers, they would have found themselves building temples and palaces for the Aztecs in Mexico City; slaves and captives who could later be sacrificed to help keep the sun in motion. America before 1492 was no simple Garden of Eden. Like the peoples and civilizations of the Old World, those in the New World had limitless capabilities for both good and evil as well. In the end, Spanish rule in Mexico would not bring relief from past wrongs, but merely replace them with others.

At the Plaza de las Tres Culturas (Plaza of the Three Cultures) in Mexico City the ruins of an Aztec pyramid stand side by side with a Catholic church and convent and a modern Mexican high-rise apartment building—each a kind of monument to its own particular place and time. The church and convent are more than four hundred years old, at one time part of a university where Franciscan monks taught the sons of the Aztec nobility. The collapse of the Aztec Empire, reads a plaque within the plaza, "was neither a triumph nor a defeat, but the painful beginning of the Mestizo people who constitute present-day Mexico." That same interpretation can be applied to the history of the Southwest as well. After the return of Spanish priests, soldiers, and colonists in 1692, Spanish and Indian culture would begin to mingle and mix. Beneath the colors of their different skins and the customs of their different cultures, these two peoples and races had, in fact, sprung from a common stock, the ancient peoples who had migrated out of Africa to populate both the Old World and the New. The close of the Ice Ages, however had separated them for more than ten thousand years. By the time they finally came face-to-face with one another they would have almost nothing in common; not only different languages and colors but different ideas of culture, religion, and society as well. The amazing thing is not that they struggled so long and so violently but that they ever learned to get along. In spite of the profound differences between them, their common humanity would slowly assert itself. In time they would become not merely enemies or strangers to one another but something far more definite—neighbors, acquaintances, friends, relatives, husbands, and wives. Unlike most of the rest of the United States, however, the Southwest was not a melting pot but a loom in which the cultures of Spain and native America were subtly woven together—each thread retaining its own distinctive color and texture but part of a common fabric.

The decades of Spanish rule that followed the Pueblo Revolt were marked not so much by domination and persecution as by the interchange of both people and ideas. For their part, the Spaniards would begin to learn the danger of their own abuses—danger that was both physical and spiritual. Security and success in the New World depended as much upon cooperation as it did domination: Push the natives too far and they could easily be thrown out. Peace was far more profitable than war when it came to running mines and haciendas. There were religious overtones to this new attitude as well. By the late 1600s an influential group of clerics in Mexico had begun to point out the hypocrisy of their missionary activities in the New World—and began to push for native rights. After 1680 the passage of the Law of the Indies would make conversion by force illegal. Other major reforms were soon underway with regard to *repartamientos* and *economiendas*, the laws that required natives to work for ranches and missions. While justice was often slow in coming, violations could be and often were punished by colonial authorities in Mexico City. There would be no more searches for Cíbola and Quivira after 1692. Instead, Spanish interests would begin to focus on the Rio Grande. Alongside the native pueblos Spanish villages began to spring up—small, tight clusters of adobe houses whose central plazas were bordered not by kivas but by churches with the wooden ends of *vigas*, or beams, protruding out from beneath the edges of their flat roofs. Beyond the walls of their villages and towns were carefully tilled fields of corn, beans, and squash. Not only did the architecture and agriculture of these Spanish settlements mirror that of their pueblo neighbors, but their social structure seemed to mirror them as well. Most Spanish villages and towns were established through a system of land grants: blocks of several thousand acres given to a group of colonists (or sometimes a wealthy or well-connected individual who then organized a group of colonists on their own) by colonial authorities and the Spanish Crown. While settlers held individual title to their homes and garden or fields, all other land within the grant—sometimes well over a hundred thousand acres—was held in common, available for grazing, hunting, and wood gathering by all within the village. Like the native pueblos, these traditional Spanish villages were not just random collections of individuals but tightly knit communities. Unlike the land rushes that would characterize later settlements in the West, the partitioning of land grants was carefully regulated: Settlers could not claim land tilled by the Indians nor could they force them from their pueblos or live among them. Within certain limits, the rights of native peoples to protect both their land and their privacy were clearly recognized by Spanish law.

While Spanish life in New Mexico came to reflect that of the pueblo peoples around them, the pueblo people in turn added bits and pieces of Spanish society and culture to their own way of life as well. Among the region's diverse pueblos, Spanish became a *lingua franca*, used by native peoples—the Hopi, Zuni, Tewa, and Jemez—to communicate with one another. In addition, Spanish authority would also force the pueblo peoples to settle their differences and disputes and live in peace with one another. Over their complex network of clans and kiva societies they also imposed a secular system of government, headed by a *gobernador*, or governor, a position still found in the pueblos today. Along the Rio Grande where Spanish influence was strongest, traditional religious beliefs became interwoven with the rites and rituals of the Catholic Church as the feast days of saints merged with those of kachinas, laying the groundwork for the rich and colorful religious life of the eastern pueblos today.

From friars and priests they learned not only the rituals of a new religion but how to read and write as well. Along with these lessons came new techniques of carving, painting, and music that they would add to their already rich artistic traditions of both pottery and architecture. New crops appeared in fields and gardens: peaches, apricots, and grapes from Spain as well as chilies and tomatoes they had carried northward from Mexico. They acquired a menagerie of animals from the Spanish as well: not only cattle, goats, and sheep but also honeybees and horses. While the silver mines of Mexico supplied Spain with wealth, the farms and ranches of New Mexico would help provide the silver mines with food and supplies, as pueblos and colonists sent wagonloads of woven rugs, pinyon nuts, brandy, and salt southward toward Chihuahua and Zacatecas.

The growing cooperation between the Rio Grande pueblos and the Spaniards antagonized their traditional enemies. Mounted on Spanish horses, the raids of Navajo, Apache, Comanche, and Ute increased in both violence and effectiveness. Native pueblos and Spanish villages became a resource to be harvested like deer or pinyon nuts. Rather than driving them apart, however, the raids simply drove them closer together. In time Pueblo warriors would march alongside Spanish soldiers in punitive raids against the Navajo and Comanche. The only respite came during the annual Taos trade fair when a temporary truce was declared while Spanish and pueblo merchants traded food and iron tools to Navajo, Apache, and Ute for blankets, furs, and even slaves. Although it is not widely known, the Southwest had a thriving slave trade in colonial times, not in black Africans but in Native Americans, captured by stronger tribes and then traded like cattle or horses. Captives and slaves had been traded among tribes long before the Spaniards arrived. Although forbidden by law, it would thrive and grow upon their arrival.

By the 1770s the attacks on pueblos and villages by Navajo, Ute, and Apache had become so debilitating that outlying areas were being abandoned. One year raiders stole so many horses that colonial authorities were forced to send off to Mexico for fifteen hundred new animals. Beyond the carefully settled world of native pueblos and Spanish villages, others were proving as adept at murder, rape, and theft as the conquistadors had before them. In 1775 visitors to Taos reported that the village resembled a walled medieval city.

Rather than a source of wealth, New Mexico became a sink for it, in need of constant assistance and supply, crippled by continual raids and brought to its knees by drought. In 1776, a few months after the Declaration of Independence had been signed in Philadelphia, two Spanish friars, Anastasio Dominguez and Silvestre Veléz de Escalante, headed north out of Santa Fe looking for a new route to Monterey on the California coast. Their expedition had been chartered to find a new route to the flourishing missions in California that avoided the territory of the feared Apache to the south. It would be the Spaniards' last great trip of discovery in the

Southwest—a group of eight men in all, traveling without maps or military escort. Their route would take them northwest along the edge of the Rockies to Mesa Verde and then northward along the flanks of the Uncompahgre Plateau into the Piceance Basin. Aided by Utes they had met along the way, they headed west across the Uinta Valley toward the Wasatch Front, crossing the mountains near Provo. After camping out for several days with Ute bands along the shores of Utah Lake they began to head south along the edge of the high plateaus, crossing the Sevier River near what is now Bryce Canyon National Park. From there they would follow the Virgin River southward to the Arizona Strip, the rugged and often forested plateau country that lies between the North Rim of the Grand Canyon and the Utah line. Working their way back east they crossed the Colorado River at Marble Canyon and then again at Glen Canyon finding a path down through the slickrock that would become known as the Crossing of the Fathers—a route that now lies submerged beneath the waters of Lake Powell. In January 1777 they rode back into Santa Fe without ever finding a route to California. In their five-month odyssey they had almost completely circumnavigated the Colorado Plateau. Unlike the conquistadors who had traveled before them, they fought no battles or wars with the natives they came across—nor were they attacked by them themselves. Much of the land they crossed would not be seen by out-siders again until John Wesley Powell's historic float trip through the canyons of the Green and Colorado rivers nearly a hundred years later.

By the early 1800s Spain's power in the New World was rapidly declining. Russian colonies had appeared in both Alaska and California. In 1803 France would sell its territorial claims along the Mississippi and the Great Plains to a newly emerging and wholly American republic known as the United States of America. A deal between Napoléon Bonaparte and Thomas Jefferson known as the Louisiana Purchase, it would give definite shape to Spanish fears of encroachment. Instead of being bordered by lands claimed by a European power whose armies were an ocean away, Spanish colonies in the region were now suddenly bordered by a young, ambitious, and rapidly expanding nation. Learning of the Lewis and Clark expedition to the Pacific Coast exploring the nation's newly purchased claims, the Spanish tried unsuccessfully to incite the Indians on the Great Plains to attack them. A few years later in 1807 they would arrest twenty-six-year-old Zebulon Pike as he and his men camped near the headwaters of the Rio Grande in southern Colorado less than two hundred miles north of Santa Fe. The year before Pike had led his group westward across the Great Plains with orders to search for the headwaters of the Arkansas River. Crossing the Rockies they had tried unsuccessfully to climb the peak outside of Colorado Springs (Pikes Peak) that now bears his name. Carted off to Santa Fe, he told the Spanish authorities that he had merely been lost. Amused by his alleged confusion, they sent him southward in chains to Chihuahua for further questioning. Eventually Pike and his men would be set free and shipped back to New Orleans, stripped of their maps and notes.

In spite of Santa Fe's magical reputation today, Pike found it a poor town of "miserable houses" whose drab walls of adobe reminded him of the flat-bottomed boats that plied the Ohio and Mississippi rivers to the east. Agricultural techniques, he said, were a hundred years behind those in the rest of the country. Rumors that the fabulous wealth of Mexican silver mines had trickled northward into Santa Fe, as he clearly saw, had no basis in fact. For their part, Pike and his men made no great impressions on the New Mexicans either. After months of cross-country travel and camp life they had looked haggard and worn. The Spaniards wondered if they were truly human and asked them if they lived in real houses or merely they were not out traveling. One thing of interest that Pike noted, however, was that cloth sold for twenty-five dollars per yard in Santa Fe while sheep brought only one dollar per head. Under Spanish control trade had been permitted with only other colonies or Spain itself. With the coming of Mexican independence in 1821, however, those controls would be relaxed. Soon traders would begin heading west from St. Louis along a route that would become known as the Santa Fe Trail with wagons crammed full of goods for the isolated but lucrative Santa Fe markets. In 1846 that stream of traders would be followed by soldiers and the start of the Mexican-American War. At the war's end two years later, New Mexico as well as Arizona, Utah, and parts of Colorado, Nevada,

Wyoming, and Oklahoma would change from Mexican to American control. For both Native Americans and Spaniards it would be like starting over again.

Meanwhile, while Spain and Mexico were fading in the early 1800s, the Navajo to the west were rising. Fearless raiders, they would acquire vast flocks of sheep and thousands of horses, growing steadily in wealth and power. The Spaniards, their fear mingled with respect, called them "Lords of the Soil."

NAVAJOS

The trail is narrow and well used. It heads almost straight down through a narrow seam in the bright red rocks of the canyon wall. Wandering through a tumbled maze of boulders and stones, the enclosing red walls of rock suddenly open up to reveal a vast expanse of green—a garden of grass and carefully planted fields lined by groves of cottonwood trees and orchards of peach trees in the midst of the desert. Hidden beneath the shade of the trees are the hogans and corrals of the Navajo, the *Dineh*, as they call themselves: the people. A broad dry wash loops and winds through the canyon floor, a river of fine white sand braided with the paths of shallow streams. The deepest carry no more than a trickle of water. Thickets of dry weeds and desert scrub growing alongside the wash are choked with debris: dried leaves and the dead branches of trees left by the flash floods that course through the canyon here, carrying the runoff of late-summer thunderstorms that drift across the surrounding desert.

In the late 1700s the canyon here was thought to be a Navajo fortress, a red-rock stronghold from which they ventured forth to plunder and raid. To the Navajo it was known simply as *Tsegi*, rock canyon, a name the Spaniards would translate as Canyon de Chelly. The fertile ground of the canyon floor has been used by the Navajo for centuries, planted with orchards of peach trees and fields of corn and beans and grazed by flocks of sheep and herds of horses. Today the canyon is both tribal land and a national monument. Cut into the flanks of the Defiance Plateau in northwestern Arizona, it lies almost directly in the midst of the sprawling Navajo reservation. Included within it are dozens of side canyons and major tributaries like Canyon del Muerto and Monument Canyon, all part of the intricate maze of converging streams and washes that flows down from the flanks of the Chuska and Lukachukai mountains not far from the New Mexico–Arizona line. Gathering together, they slice through the red rocks of the desert to carve out Canyon de Chelly. While the rock walls near the canyon mouth are no more than thirty feet high, in the space of a few miles they rise to nearly a thousand feet high, towering above the flat, narrow reach of the canyon floor.

As we walk along the canyon walls toward the junction of Canyon del Muerto, Perry Yazzie points out pictographs and petroglyphs on the canyon walls: images of antelope and deer, the dancing shape of yei figures and humpbacked flute players. Some are Navajo. Others are the Pueblo. A few are Spanish—the names of passing soldiers and priests. The day is warm and bright, a Sunday afternoon in early September. As we head up Chinle Wash toward the mouth of Canyon del Muerto, we hang close to the canyon walls, walking through groves of Russian olives and towering cottonwood trees. Puffy white clouds float through the sky overhead. Out in the center of the canyon traffic is heavy: tourists in jeeps and four-wheel drives with native guides and Navajo families in pickup trucks loaded with groceries and children. Although the land here is a park to outsiders, to the Navajo it is a place to live and work. Yazzie points out the hogans, fields, and corrals of relatives tucked away in alcoves and side canyons behind a green screen of trees. Like most canyon residents, Yazzie says, his own family lives here only seasonally, tending gardens of corn, beans, squash, and melons in the summer and moving out into the desert flats outside of Chinle in the winter. Near the canyon mouth, he adds, the sandy floor of the wash makes a good place to have a barbecue and play volleyball. The tourists, he says, are always kind of surprised. Life here may be traditional, but it is not lost in time.

As we round a bend in the curving path of the canyon, the ruins of an ancient pueblo suddenly appear in the canyon wall. More than a hundred feet above the canyon floor, they look almost as if they had been carved right out of the rock, a tight, narrow cluster of apartments and granaries clinging to the side of the cliffs with black, blank doorways and windows overlooking the canyon below. Seven hundred years ago the canyon here was home to the Anasazi, relatives of the same peoples who had settled Mesa Verde and Chaco Canyon to the east. Here in the cliffs of Canyon de Chelly they built more than a dozen small villages, only to abandon the area by 1300 A.D. The land belongs to those who live on it. Nearly two hundred years after these prehistoric pueblo peoples had abandoned it, the Navajo would arrive and make this fertile and beautiful canyon their own.

The Navajo had no contact with these prehistoric pueblo peoples, nor were they relatives. They called them *Anasazi*, a word that means the ancient ones, but also the ancient enemy, the ancestors of those who are our enemies. Believing them to be contaminated with the spirits of the dead, the Navajo left their abandoned cliff dwellings almost completely untouched and settled on the canyon floor to raise their crops and tend their flocks. Mistakenly believing the Navajo word *Anasazi* meant simply "the Ancient Ones," American archaeologists working in the Southwest would begin applying it to the prehistoric pueblo peoples of the Four Corners region.

Although they live side by side, the peoples and tribes of the Southwest have diverse histories and origins. Unlike the pueblo people who preceded them, genetic and linguistic data suggests that the Navajo were not descendants of the Amerind peoples who migrated to North America from Asia some fifteen to twenty thousand years ago but more recent arrivals—descendants of the Na-dene peoples who migrated to North America some eight to ten thousand years ago. Settling in the boreal forests of northern Canada, they would become the Athabascan people of the Canadian and Alaskan interior. Some one thousand years ago two groups would split off of this main stock and begin heading south—one toward the Pacific Coast, where they would become the Coastal Indians of southeast Alaska, British Columbia, and the Pacific Northwest; another heading toward the Four Corners region, where they would become the Navajo and Apache. Following the front range of the Rockies to their end near Albuquerque and Santa Fe, they then turned west to spread out through northern New Mexico and southern Colorado and into the canyons and plateaus of the Four Corners region.

Traditional Navajo, of course, place no more faith in these scientific theories of their origins by anthropologists and archaeologists than fundamentalist Christians place in biological theories of evolution or geologic theories of a 4.5-billion-year-old earth. Like the Pueblo and the Ute, they believe that they have been here in the Southwest since the beginning of time. "Did you know that there are people like us in Alaska and Canada?" one Navajo told me at an elaborate dinner at the home of his Zuni in-laws on Shalako night. "Their language is almost exactly like ours. They must have been taken away as slaves or something a long time ago," he said.

For the Navajo the mountains and deserts of the Four Corners region would become both a homeland and a holy land, a place where not only their ancestors once walked the earth but the gods as well—Changing Woman, Talking God, and Monster Slayer.

After migrating through a succession of colored worlds below—black, blue and yellow—the Holy People would emerge into the Fifth and present world on a small island in the midst of a large lake. While the pueblo peoples like the Hopi and Zuni trace the emergence of the kachina spirits to a series of sacred lakes on the Mogollon Rim of central Arizona, the Navajo believe that the place of emergence of their own ancestors lies on top of Huerfano Mesa in northwestern New Mexico or somewhere farther north amid the San Juan Mountains of southwestern Colorado. Its exact location is uncertain. There is no Navajo equivalent of the *Torah* or *Bible*, no official written account of their early history. Traditional stories are passed down orally from generation to generation and committed to memory, but the fine details of the elaborate stories of the Navajo creation myth and the Navajo way vary from clan to clan and region to region.

Once on the surface of this new world the Holy People and the gods would begin to prepare it and make it safe for the Navajo whom they would later create. First Man and First Woman would mark the cardinal directions with four sacred mountains, building them from materials they had carried with them from the Fourth World below and fastening them to the ground with sunbeams, rainbows, and bolts of lightning: *Sisnaajini*, Sierra Blanca Peak in the east; *Dibé nitsaa*, Mount Hesperus in the north; *Dook 'o' ooslid*, San Francisco Peak in the west; and *Tsoodizil*, or Mount Taylor in the south. At night they decorated the sky above with stars and constellations, fashioning them out of chips of glittering mica and jewels. While

First Man and First Woman would labor to give the earth shape and beauty, other gods like the Hero Twins would arm themselves with weapons from their father, the sun, to destroy the monsters who preyed upon the earth's first people: giants like Yeitso who lived near Mount Taylor and whose footsteps caused the earth to tremble and shake and winged monsters like the Tsenahale who used the solitary volcanic neck of Shiprock, known as Winged Rock, *Tsé bit'a'í* among the Navajo, as an aerie; and countless others who inhabited the earth's first canyons and mountains and lurked in the abandoned ruins of pueblos. In the legends of the Navajo the lava fields of El Malpais are not the blood of a blind kachina but the blood of a giant killed by the Hero Twins. The fossils of dinosaurs are not the remains of ancient lizards and reptiles but the bones of monsters slain by the gods.

The Hero Twins' epic battles with the monsters would conclude with a four-day storm of hail, high wind, and heavy rain that all but wiped out the scattered evil spirits that remained and changed the face of the land forever. Before it, the Fifth World had been rich and green, covered by thick soils and dense forests. After it, the wind and rain would leave behind only desert: carving the surface of the earth with deep canyons and leaving behind only solitary pillars and spires of rock. Once the world was saved, the gods and the Holy People would gather together to create the first *Dineh*, the five-fingered earth surface people, from ears of corn. Nílchi'i the wind would give them the breath of life. Look closely at the patterns of whorls that cover your fingertips, they say, and you can see the trail of the wind.

Beyond this point the facts of legend and those of history begin to merge. After their appearance in the Fifth World (unlike the pueblo people, traditional Navajo believe that this is the Fifth World, most of their traditional stories containing not four worlds but five), the scattered clans of the Navajo traveled from place to place, finally gathering together in northwestern New Mexico in the drainages of Gobernador and Largo Canyons, not far from Chaco Canyon. While the Navajo claim to have arrived here while the Anasazi were still building their pueblos, archaeological evidence puts their arrival somewhat later: in the late 1400s or early 1500s, only a few decades before the arrival of the Spanish. As they had for the Anasazi before them, the barren deserts of the Four Corners region would allow the Navajo to live almost undisturbed, gathering in strength and number as they borrowed bits and pieces of ideas and culture from those around them to build a unique way of life all their own.

The relationship between Navajo and Pueblo is complex in both history and legend. Throughout the elaborate cycle of stories that make up the Navajo creation myth, the ancestors of the pueblo people appear repeatedly, living alongside them in the early days of the Fifth World, both a companion and a counterpoint to the lives and ways of the Navajo. Archaeologists believe that the first Navajo to reach the Four Corners region were simple hunters and gatherers whose nomadic lifestyle resembled that of the Paleolithic hunters and gatherers who had preceded both the pueblo people and the Navajo. Like them, they lived on foot and traveled constantly. When the Navajo first arrived in the southwest, the crowded pueblos of the Anasazi with their plazas and kivas must have seemed almost unreal.

From the pueblo people they learned how to plant fields of corn, which they called the food of the strangers, as well as beans and squash. Watching their elaborate kachina ceremonies, they began to develop their own elaborate rituals as well: songs and curing ceremonies centered not around masked dancers but elaborate sand paintings and chants. Not only did they borrow freely from the pueblo peoples, they stole freely from them as well. While the pueblos with their adobe walls were almost impenetrable, the crops in the surrounding fields as well as those working among them were easy prey. From time to time the pueblo people mounted their own raids upon the Navajo in retaliation. Two very different peoples, they became both friends and enemies.

Changes came slowly. When the first Spaniards arrived, the Navajo were still a small, weak tribe, moving from place to place in the wilderness that lay beyond the more settled world of Spanish villages and Indian pueblos. No longer purely nomadic, they lived near their planted fields in pithouselike hogans—round buildings whose roofs were covered with piles of stick and brush. While the Spaniards called the descendants of the Anasazi they encountered *Pueblos*—Spanish for village or town—the Dine would become known as the *Apaches de Nabaju*, the wanderers of the cultivated fields.

While Spanish diseases and Spanish rule decimated the Pueblos, the Navajo thrived. Growing in strength and number, they stepped up their raids, stealing not only herds of sheep but the herders to go with them as well. While their Apache relatives joined in the Pueblo Revolt to help drive the Spaniards from New Mexico, most Navajo preferred to sit on the sidelines, collecting whatever came their way in the way of horses, sheep, and refugees. "War brings rain and blessings," a traditional saying among the Navajo claims. "Because of it flowers which have become beautiful exist. Because of it rain exists." While the wind would give life to the Navajo, flocks of Spanish sheep and herds of Spanish horses would transform them into a rich and powerful tribe. They would grow dramatically in numbers as well. The return of Spanish forces in 1692 would swell their ranks with hundreds and perhaps thousands of refugees, as dozens of pueblos were abandoned and their residents moved in with the Navajo.

While the women learned weaving and pottery from the refugees, imitating them point by point, the men learned the details of Pueblo ceremonies and combined them with their own beliefs. They learned to build in stone as well, some groups settling for a time in loosely organized pueblos of their own with clusters of rooms and watchtowers. While Navajo and Pueblo refugees lived together, they also went southward in a series of punishing raids on the pueblos that had submitted to Spanish rule and the Spanish villages themselves. Eventually the Navajo would be forced out of northern New Mexico, not by the Spanish but by an alliance between the Ute in the mountains to the north and the Comanche of the Great Plains to the east. Driven from New Mexico, they would begin to settle in the red-rock deserts of northeastern Arizona in the late 1700s centered around Canyon de Chelly. While the Chuska and Lukachukai mountains were as rich and green as the Rockies and Sangre de Cristos they had been forced to flee in New Mexico, the slickrock deserts that spread out from their western flanks were laced with canyons where their flocks and fields could remain hidden and safe. Secure in their newfound home, the Navajo stepped up their raids. The Spaniards and Pueblos mounted retaliatory raids of their own, and for decades the various shifting sides swapped raids, murders, and slaves.

The conflict came to a head in 1803 when the Spaniards enlisted the help of a professional Indian fighter from Mexico by the name of

Antonio Narabona. With an army of three hundred Sonoran troops and auxiliaries drawn from the ranks of the Opata Indians of northern Mexico, Narabona headed northward to hunt the Navajo down. Led by Zuni guides, Narabona and his troops entered the northern arm of Canyon de Chelly in January 1804. After hiding the elderly and their women and children in a cave some six hundred feet above the canyon floor, Navajo warriors moved to the canyon rim to watch the Spaniards and their Indian auxiliaries move through the canyon below. They had used the hiding place before to escape bands of marauding Ute. This time, however, it would turn into a death trap. As the soldiers and their auxiliaries passed by, a women in the cave who had served as a slave to the Spanish began taunting them with curses. The Spanish answered with bullets. While snipers shot down those whose heads appeared above the rocks, others sent a shower of bullets ricocheting off the cave's roof into those who lay hidden within. The next morning they scaled the cliffs themselves and finished off the survivors with clubs and swords. In all more than a hundred died. Only one, an old man left for dead, would survive to tell the others what had happened. Their hiding place would become known as Massacre Cave; the northern arm of Canyon de Chelly as Canyon del Muerto—Canyon of the Dead.

After the bloody attack, an uneasy peace prevailed. Eventually the Spaniards would be driven out of the Southwest for good, not by the Indians to the north but by internal troubles to the south, a bloody war of independence that would transform the former Spanish colonies of the region into an independent nation. By 1821 New Mexico was no longer part of Spain but of Mexico. Little changed for the Navajo: The Spaniards among them were now Mexicans. With the Spaniards gone and the Mexicans still struggling for control, the Navajo would rise again. For the next twenty years the rains were plentiful, blessing both the fields of the Navajo and those of the pueblos and ranches around them. The raids would continue. While the elders grew rich and pleaded for peace, the young men still had fortunes to make—tending herds of their own and stealing others. By the time the Americans arrived they would all but rule the region.

After a two-year war between Mexico and the United States in 1848, the Southwest would pass from Mexican to American hands.

Under the terms of the Treaty of Guadalupe, the United States would pay the Mexican government some $18 million for war damages and gain control of the land that now makes up New Mexico, Arizona, California, Nevada, Colorado, and parts of Utah. Coupled with the loss of Texas in 1845, by the war's end Mexico had lost nearly half of its territory to the United States. Santa Fe would surrender almost without firing a single shot. The coming of the Americans would change the world of the Navajo almost completely, although they did not realize it at the time. The Southwest was no longer part of a remote frontier but caught in the midst of a rapidly growing nation; claimed by a new people, very different from the Spaniards, who believed they had been chosen by God to build a nation whose borders reached all the way from the Atlantic to the Pacific.

The Americans, however, had inherited not only the vast reach of land here but a tangle of conflicts several centuries old. While the Spaniards had eventually forced the pueblo peoples to live in peace with one another, the Americans would find themselves trying to mediate what was often a three-sided struggle among the Navajo, Pueblo, and Mexicans. While U.S. leaders cautioned them to keep the peace, the Navajo pleaded for a chance to settle their differences unhindered. When told by the commander of Fort Wingate not far from Mount Taylor that the Mexicans were now U.S. citizens and entitled to protection, the Navajo were incredulous. A young Navajo leader by the name of Long Earrings tried to explain their frustration: "Americans!" he said. "You have a strange cause of war against the Navajo. We have waged war against the New Mexicans for several years. We have plundered their villages and killed many of their people and made many prisoners. You have lately commenced a war against the same people. You are powerful. You have great guns and many brave soldiers. You have therefore conquered them, the very thing we have been attempting to do for so many years. You now turn upon us for attempting to do what you have done yourselves."

While Navajo raiding had decimated the flocks and fields of the Mexicans, retaliatory slave raids by the Mexicans and others had in turn captured hundreds of Navajo. Although they soon made plain their intent to curb the raids on the Mexican ranches and Indian pueblos now under their control, U.S. troops and government officials proved unable or unwilling to force the Mexicans to return their Navajo slaves. Treaties were signed and agreements were made, but neither the treaties nor the peace lasted for very long. Efforts at mediation predictably ended in failure. The problems were complex. "Perhaps things might have been better had the Army of Occupation been sufficiently large and had its soldiers been well drilled, well supplied, and perfectly behaved," Ruth Underhill suggests in her book *The Navajos.* "Also had its officers been statesmen with a knowledge of Spanish, as well as economics, sociology, and anthropology."

To hold the Navajo and other tribes in check, the United States set up a system of forts and Indian agents in the West to distribute treaty goods: payments of food and clothing in exchange for peace. While old resentments and abuses simmered beneath the surface, the raids slowed. In northern New Mexico and southern Colorado, Spanish settlements that had been abandoned decades before were resettled while peace prevailed. With the onset of the Civil War, however, it would all come unraveled. In 1862 Confederate troops would briefly seize and hold Santa Fe. As fighting intensified in the East, troops were recalled from the Southwest and the forts were left abandoned. The Navajo cared little as to why the soldiers had left. Seeing themselves free to settle old scores, they went back on the warpath and launched a series of crippling raids against both the Mexicans and the Pueblos. In the midst of a war that would claim more than half a million and nearly destroy the Union, U.S. forces had no time or sympathy for instability or unrest in the West. In 1863 while the Union army was handing the Confederates a bloody defeat at Gettysburg, a detachment of the U.S. Army under the command of Christopher "Kit" Carson set out to silence the Navajo for good.

While the Navajo numbered more than ten thousand, Carson had an "army" of no more than seven hundred poorly trained troops who were short on horses, ammo, and supplies—and the Navajo they were ordered to find and destroy were spread across several million acres. At Taos, Carson had served as an Indian agent to the Ute. An experienced frontiersman and scout, he did not waste time pursuing individual bands of Navajo as his predecessors had done before him, but hunted crops instead, destroying fields of corn and burning stores of food. In a single raid they seized more than seventy-five thousand pounds of wheat. While Carson rode with his soldiers, he had little need for an army. Along with them were hundreds of Ute, Zuni, Hopi, and Mexicans, a willing band of Navajo-haters ready and armed for war. When money ran short, the Isleta pueblo loaned $60,000 to the U.S. Army to enable the soldiers to receive their back pay, ensuring that they would stay in the area to finally drive the Navajo from their midst. Bounties were offered: twenty dollars for every Navajo horse and one dollar for every Navajo sheep. Captives they were free to keep as slaves, and hundreds of Navajo disappeared for good among the Hopi, Ute, and Mexicans.

In six months Carson's scorched-earth tactics would bring the Navajo to their knees. In January 1864 they marched through Canyon de Chelly chopping down peach trees and burning hogans and corrals. That summer they had watched the peaches ripen and fall, daring only to pick them at night, using bunches of weeds to brush away their tracks. The destruction of Canyon de Chelly broke the Navajo's back. Afterward Carson no longer had to go out and look for the Navajo. Starving to death and dying of exposure, they came to him, surrendering in small bands and groups and asking for peace.

The Navajo, however, were not only to be brought under heel but sent off into exile. Their new home as envisioned by federal officials in Washington would not be in the Southwest but out on the Great Plains, joining the Apache (their relatives and sometimes enemies), who had been sent there before them on a spread of some forty square miles of high plains on the alkaline banks of the Pecos River that would become known as Bosque Redondo. The first caravan started east from Fort Defiance near Canyon de Chelly in April 1864 with some twenty-four hundred Navajo. After stopping for a few days outside of Albuquerque, which the Navajo called "Place of the Bells," they marched eastward beyond the mountains and out into the plains, roughly following the route Coronado had taken some three hundred years before. Except for the most wide-ranging buffalo hunters among them, it was the first time any of them had ever seen a landscape so totally devoid of both rocks and trees. They made the three-hundred-mile trip in just fifteen days.

Over the next few months more than eleven thousand Navajo

would find their way to Bosque Redondo. Not all, however, would be driven off into exile. Some four thousand would remain behind, hiding out amid the canyons and plateaus or seeking refuge among their pueblo neighbors as the pueblo people themselves had done after their failed revolt more than a century and a half before. Others, it is said, escaped through a rocky tunnel under the San Juan River and fled to the north.

Those at Bosque Redondo lived on the edge of starvation for more than four years. Preoccupied with the close of the Civil War and Reconstruction, federal authorities had little interest in the plight of a few thousand Navajo. In a matter of months the thin stands of trees that lined the riverbanks were cut completely away and the Navajo were traveling fifteen and twenty miles to find wood, uprooting clumps of mesquite and gathering their roots to burn as firewood. With six thousand acres of irrigated farmland, government officials had hoped to turn the Navajo and Apache into farmers and have them raise their own food. The waters of the Pecos, however, were too alkaline for farming. Hail and army worms took care of the rest.

With the Navajo starving and money in short supply, U.S. Army troops in New Mexico went on half rations to keep the Navajo alive. Supplies were minimal: flour, coffee, and sometimes bacon. When printed ration tickets were introduced, the Navajo quickly learned to forge their own, increasing their supply of food. Later these were replaced by stamped metal tickets that they learned to duplicate as well. Troops soon began allowing the men to travel out from the camp and hunt. There was, however, no escaping the squalid conditions of the camp. All in all as many as a fourth of those who had been herded off to Bosque Redondo would die in captivity. Despair was no less deadly than disease: Outside the boundaries of their sacred mountains, their medicine men refused to perform their elaborate ceremonies. The old ways had lost their power. In captivity they found no relief from their traditional enemies either but were raided by roving bands of Comanche and Mexicans. Group by group and family by family, the Navajo began trickling away from the camp and heading back to the Four Sacred Mountains to find their relatives who had stayed behind.

Bosque Redondo was a failure. Soon even the Americans were

saying it and looking for a new place to send the Navajo. Plans were laid to send them to more fertile ground in Oklahoma and Arkansas, but the Navajo wanted nothing of it. Negotiating with the officials who would decide his tribe's fate, the Navajo leader Barboncito, stories say, pulled a knife from his moccasins and threw it to the floor, telling them: "If you wish to send my people from their home, first take this knife and kill me." The Americans relented. Two weeks after signing a peace treaty that pledged them to live in peace with those around them, the Navajo began heading west toward the Four Corners area and Canyon de Chelly, the land that lay between their Four Sacred Mountains.

At eighty-three Luke Barney can still split his own firewood. After driving up the dirt road that leads to his house just off Highway 666 near Tohatchi, New Mexico, I found him outside chopping kindling with a long-handled axe. The logs were six and eight feet long, stacked up on end and leaning against one another like a bonfire waiting to be lit.

We spent the afternoon talking at a table inside the house while his wife rested in an adjoining room, talking about growing up on the Navajo reservation. "I don't remember much until I was about five years old," he said. "I learned most things from my mother. She taught me how to take care of things: sheep, cattle, things like that. In the springtime we planted and worked in the fields, all the way through the summer and into the fall. In October we harvested the crops, storing part of it for the cattle in wintertime."

By the time he was fourteen both his father and mother had died. Barney and his two brothers went to live with their aunt until they were old enough to take care of themselves. He started school in nearby Fort Defiance in 1923 and later went on to the Indian Boarding School at Phoenix, Arizona, leaving school in 1928 when one of his brothers asked him to come home to help him tend their growing herds of sheep, cattle, and horses. Barney's brother would eventually marry and start a family of his own. Later he would do the same as well. He and his wife would eventually raise fourteen children—ten girls and four boys. While many still live right nearby, others are scattered as far away as Page, Arizona, and Washington, D.C.

Barney himself has spent most of his life right here around the edges of the Chuska and Lukachukai mountains, farming and raising cattle and sheep. The house and land here, he says, belong to his wife's family. Tending sheep has a regular rhythm to it, and the place here near Tohatchi serves as their winter range. "We usually come back to this place in the fall around November 20 and stay until sometime in February," he explained. "In March or April we have lambs and after two or three weeks we go back to Narrow Creek outside of Crystal, New Mexico, again and stay there all summer. That's the way our season is." The land over toward Crystal on the other side of the mountains was settled by his mother's family after they returned from several years of confinement at Bosque Redondo. The land up in the mountains is rich and green. "He'll probably see if you can drive him up there if the weather's good," his daughter had told me when giving me directions the night before. "He loves it up there." Taking out a piece of paper, he draws out a little map—the path of the creek through the land, the position of the corrals and the hogan, and the layout of their fields nearby. As we sit by the warm woodstove, the temperature outside is rapidly falling below freezing and the wind is blowing at more than thirty miles per hour. In a few days the mountains will be covered with snow.

Today Barney's children and their families take care of the sheep and the farming and also hold down jobs in Window Rock and Gallup. "I can't do the work like I used to when I was young," he said. "I leave the whole thing up to my children. I'm eighty-three. That's quite a while."

According to government records some 8,121 Navajo would return from Bosque Redondo in 1868, bringing nearly two thousand sheep and goats with them, survivors like themselves of four years of exile. The treaty that enabled them to return to their homeland also promised livestock, and in 1869, a little more than a year after their return, fourteen thousand sheep and a thousand goats would be brought to the Navajo. Out of those few thousand animals the Navajo would build up a herd of more than a million sheep by the turn of the century and increase their own ranks to more than twenty thousand. While other tribes would see both their numbers and lands shrink under American control, both the numbers of the Navajo and the size of their reservation have increased with time. Today the Navajo reservation sprawls over more than 17 million acres, while the Navajo themselves number more than two hundred thousand.

After Bosque Redondo there would be no more wars with the U.S. troops, and eventually the Navajo would come to fight alongside them as well. In the Second World War, U.S. armed forces in the Pacific used Navajo "code talkers" to communicate with one another. Adept at cracking codes, Japanese Intelligence never managed to unravel the intricacies of the Navajo language that they so carefully monitored. Others would serve with distinction in Korea, Vietnam, and the Persian Gulf along with thousands of members of other native tribes. It would take Natives and Americans the better part of a century to learn to live together—a painful process that still continues today. Out of the ruins of Bosque Redondo the Navajo would rise again.

While the Spanish brought new tools and crops to the pueblos and in some cases a new religion as well, the pueblo people remained tied to their traditional villages and carefully cultivated fields. By contrast, although there was little direct contact between the Spanish and the Navajo, the tools and animals the Spaniards brought with them would reshape the world of the Navajo almost completely. Acquiring flocks of sheep from the Spaniards, the Navajo would develop a pastoral lifestyle all their own. Neither entirely settled nor completely nomadic, they moved from season to season and camp to camp with their flocks. But while their new way of life centered around flocks of Spanish sheep, it was nothing at all like the culture the Spanish had brought along with them. The history of Navajo culture is not a simple replacement of native values by European ones, but a careful selection and re-creation of them— something the Navajo had done and would continue to do throughout their history. Rather than simply borrow ideas, they remade them—like someone taking an old blanket and unravelling the threads and then using them to weave a new and entirely different one. From the pueblo people they acquired corn, but not their tightly packed villages and towns. From the Spaniards they acquired sheep, but not their religion. In time other bits and pieces of Spanish culture would find their way into the Navajo world: indigo dye and red bayeta for cloths and blankets, and also silversmithing—which the Spaniards, in turn, had learned from the Moors during their own long occupation before 1492.

Spanish influence on Native America reached far outside the Southwest as well. Out on the Great Plains, the arrival of Spanish horses would change the balance of power almost completely, trans-forming small, weak tribes like the Comanche, Sioux, and Kiowa into strong nations with territories that spanned hundreds of miles. As their power increased, so too did the complexity of their art and culture. The same was true of the Ute and Navajo in the Southwest as well. While sheep had made the Navajo wealthy, the horse would make them mobile, giving them the power to hunt, travel, and raid over literally hundreds of miles. While the Spanish and other Europeans took a great deal out of North America, they also brought a great deal with them. The Native America encountered by American pioneers in the early 1800s and in the years that followed the Civil War bore little relationship to the one that had greeted Spanish conquistadors in the early 1500s.

Reservation life brought its own changes as well. Indian agents and administrators brought in blacksmiths and silversmiths to teach the Navajo new crafts, which they quickly adopted and refined. Private trading posts scattered across the reservation brought con-tact with the outside world as well: calico, velvet, canned peaches, and coffee. Finding a new market for Navajo rugs and blankets in the East, they would begin buying their handwoven blankets and urge them to shift to rugs, fostering the rise of what would become a rich and valuable artistic tradition. In time their work, prized by collectors and galleries around the world, would come to over-shadow that of the pueblo people who had first taught them how to weave.

Raiders and invaders from every direction had reshaped the face of Europe with war and trade for more than one thousand years. The roots of so-called traditional culture in both the Old World and the New come from many sources. In the New World those same Europeans would continue that long history of trade and war, reshaping the lives of those they came in contact with. It has become fashionable today to portray Native Americans as nothing more than victims. While there is no doubt that Native Americans often suffered mightily, to see their history as simply one of submission is an insult to both their achievements and their memory—a facile manipulation of the facts that ignores the complexity of the relationship between immigrant and native; the struggle between old and new. Cultures and peoples in the New World were not fixed and unchanging before their contact with Europe. Paleolithic hunters would become pueblo farmers. Athabascans from the forests of northern Canada would appear in the deserts of the Southwest. With the arrival of Europeans and Americans, however, the pace of that change would accelerate dramatically. Rather than changing in response to decisions of their own, Native Americans would increasingly find themselves forced to change in response to the decisions of others. In time their lives would become twined together. Five hundred years after their arrival, native and immigrant are still struggling to understand one another.

BOOK FIVE

West

THE IDEA of the landscape as a text has long been part of the American consciousness. To the seventeenth-century Puritan imagination it was sacred, revealing God's designs to the elect. To nineteenth-century travelers in the West it was often a text in Manifest Destiny, or in natural law, or in the aesthetics of the picturesque and the sublime. To one whose sense of things derives from a particular place, the shape of a particular horizon that constitutes the only proper edge of the sky, it may be highly personal, a diary, even a confession. But all texts require interpretation and tend toward ambiguity when closely scrutinized. Moreover, they always reflect the assumption the reader brings to them. Thus, the process of reading landscape is in large part an experiment in self-discovery, and the process of writing an exercise in self-revelation.

EDWARD A. GEARY
The Proper Edge of the Sky

S E V E N T E E N

Rising out of the Desert

Heading east on Interstate 40 from California, the Colorado Plateau seems to rise up out of the ground like a high green wall. Its edge in northwestern Arizona is marked by the four-thousand-foot-high wall of the Grand Wash Cliffs. They appear east of Kingman, running north through the desert toward Utah. The cliffs mark the end of the Grand Canyon as well. Farther north the Colorado River slices through the Grand Wash Cliffs leaving the thin narrow world of the canyon for the broad open reach of Lake Mead. After tumbling through the turbines of Hoover Dam to generate power for sprawling, dry cities like Las Vegas, Phoenix, and Los Angeles, the river heads south through the low, hot deserts that straddle the California-Arizona line, heading toward Mexico and the Gulf of California, where it finally meets the sea.

In the base of the cliffs you can see rocks of the same age and appearance as those that line the walls of the inner gorge of the Grand Canyon farther east. Here, however, it is not erosion that has laid these rocks out in plain view but uplift. A deep fault known as the Grand Wash Fault runs along the base of the cliffs. Rocks to the east have been uplifted by more than a mile.

Beyond the reach of the cliffs the highway seems to climb up the edge of the plateau through a series of stairsteps—a progression of flat-topped mesas that leads steadily upward. As the land rises, the Joshua trees and ocotillo of the desert slowly give way to woodlands of pinyon and juniper. The air is suddenly crisp and cool. As the land climbs above seven thousand feet, forests of ponderosa pine and Douglas fir appear. Meadows of rich green grass lie scattered among the trees like a chain of lakes and pools. The highway skirts the southern edge of the Coconino Plateau, a broad, high reach of ground whose gently rolling surface is broken only by the scattered shapes of cinder cones and the high, solitary rise of the San Francisco Peaks. East of Flagstaff the plateau drops off into the Painted Desert nearly a half mile below. To the north, however, the forests seem to go on forever.

Here in the west the edge of the Colorado Plateau is defined by a line of high plateaus. They run northward from Arizona like a high green spine: the Shivwits, Uinkaret, Sevier, and Wasatch plateaus—through Utah all the way to the edge of the Uinta Mountains. In places their flat-topped crests are more than two miles high, rising up more than six thousand feet from the deserts below.

On a still, clear day you can stand on top of the Wasatch Plateau in central Utah and see clear across the Colorado Plateau, all the way to the San Juan Mountains in Colorado that mark the western edge of the Rockies. In between are broad desert valleys and deep canyons carved into layers of red and tan-colored rock. The flat sweep of the ground is broken by the rectangular, sharp-sided shapes of mesas and buttes, a landscape of levels that stretches as far as the eye can see. Here and there solitary peaks rise up out of the ground like giant pyramids: the La Sal, Abajo, and Henry mountains. In the foreground you can see the wavelike rise of the San Rafael Swell, a wall of solid rock that runs across the desert for more than forty miles.

The mountains are not part of the Rockies farther west nor volcanoes like the San Francisco Peaks or Mount Taylor farther south but something completely different. While eruptions and flows of volcanic rock laced the edges of the Colorado Plateau with cinder cones and lava fields, here near its center they barely broke the surface. Instead of erupting, molten layers of rock below rose up slowly, pushing the layers of rock above them upward like a piston, leaving behind ranges of solitary peaks known to geologists as laccoliths. They include not only the Abajos, La Sals, and Henrys but also

Navajo Mountain near the northern edge of Lake Powell and Sleeping Ute Mountain near the Four Corners. Elsewhere there were no volcanic rocks at all, merely bursts of superheated steam from pools of magma below that blew their way to the surface in conduits or pipes known as diatremes. At places like Shiprock and the Hopi Buttes these diatremes not only altered the rocks around them but brought blocks of exotic rocks to the surface from the earth's mantle some thirty miles below.

To the west the view is entirely different. The landscape is not broken by the rectangular shapes of plateaus and mesas but serrated by ridge after ridge of high plateaus: the Pavant, House, Confusion, and Indian Peak ranges that corrugate the Basin and Range country of western Utah. At times you can see all the way to the Snake Range in eastern Nevada. Fifteen to 20 million years ago the land here began to stretch and thin. As the earth's surface began to spread, its crust broke into blocks that tilted and sank, like blocks of ice on a half-frozen river, their upturned edges forming mountains. Like the rise of the Rockies to the east, the faulting and folding that created this new belt of mountains to the west left the Colorado Plateau region almost untouched. Instead of breaking up, the plateau began to rise like a ball of freshly kneaded dough. Sometime between 20 and 5 million years ago the Colorado Plateau was uplifted by more than a mile—not in bits and pieces but as an almost solid block, as if the whole region had been suddenly jacked up into the sky. Neither the isolation and stability from the disturbances that surrounded it nor the forces that would later thrust it up toward the sky are clearly understood. With the region's uplift, however, the creation of the plateau would come full circle. Built up over the course of several hundred million years, it would now begin to slowly wear away. As the land rose up, the rivers would begin to cut down, carving its surface with deep canyons and cliffs to create a landscape unlike any other in the Intermountain West.

Although they define the edge of the Colorado Plateau, the high plateaus of central Utah and northwestern Arizona seem like a world apart. While deserts of sage and grass and dry woodlands of pinyon and juniper cover the flanks and crests of the lower mesas and plateaus, to the west the tops of high plateaus are rich and green. In late June on top of high places like the Wasatch Plateau, drifts of snow still cover the ground, broken by meadows of green grass and alpine flowers: carpets of yellow avalanche lilies and white spring beauties. Below the reach of these high meadows are forests of aspen, spruce, and fir. This too is part of the Colorado Plateau. Look closely in the side of the high cliffs that mark the sides of the plateau and you will see the same tan and brown sandstones that make up Mesa Verde to the east. Above them are brightly colored sandstones in shades of pink, white, and orange—the same deposits that make up the spires and hoodoos of Bryce Canyon farther south. The only difference here is one of altitude. While the deserts below typically receive no more than ten inches of rain per year, the tops of these high plateaus are typically soaked with more than forty. Down in the desert at five thousand feet, summer temperatures frequently climb above a hundred degrees. Here on the plateau above ten thousand feet they rarely climb above sixty. While extreme faulting has lifted these rocks high into the sky, they are capped by thin layers of lava that keep them from eroding away.

You can find the same cool, green landscape all around the plateau wherever the land rises high enough: the Uncompahgre Plateau in Colorado, Mount Taylor in New Mexico, the Kaibab Plateau in Arizona, the Abajos in eastern Utah. Fifteen thousand years ago during the Ice Ages forests of spruce and fir like these covered the desert below, while the plateaus and peaks themselves were covered with permanent fields of snow and ice. Glaciers and streams carried boulders and stones from their summits down into the valleys below. To the east in the Basin Range, the narrow valleys were filled with lakes. Eight to ten thousand years ago when the climate began to warm, desert plants and grasses from the south began to drift northward, replacing the trees in the valleys below. As the glaciers retreated, the pines and firs began to climb up into the peaks and high plateaus. Today they exist almost like islands, an isolated world of green floating several thousand feet above the desert, an alpine world of flowers, snow, and trees.

While Spanish influence in the Southwest would come to center around the Rio Grande to the east, Mormon influence in the region would come to center around the high plateaus to the west. Fleeing religious persecution in the eastern United States, the first Mormon pioneers reached Utah's Salt Lake Valley in 1846. They settled at the base of the Wasatch Front, a line of high, jagged peaks that runs north toward Idaho and Wyoming, replacing the high plateaus that mark the edge of the Colorado Plateau farther south. Others had called the region worthless, but the Mormons saw the land around them in biblical terms. The Great Salt Lake was a vast Dead Sea; the deserts beyond it as barren as those of Israel. The salty stream that flowed through the center of their valley from Utah Lake to the south they called the River Jordan. With faith and hand work they believed they could make the desert bloom like a rose.

Unlike other pioneers, the Mormons did not continue heading west to the Pacific where the land was rich but stayed in the desert, hoping its ruggedness and isolation would shield them from the world outside. Traveling by foot and on wagon trains (the railroad would not reach Salt Lake until the late 1860s), their ranks grew by several thousand each year. As the population of their New Jerusalem began to swell, new pioneers pushed outward from Salt Lake, some heading north along the mountains while others pushed south along the edge of the high plateaus. Sites were carefully scouted and selected, located on the sides of streams and creeks that flowed out of the high plateaus behind them. Perhaps more than anyone else before or since in the West, early Mormon pioneers understood the link between mountain and desert: that snow and rain falling in the high peaks and high plateaus around them were the key to their survival. They stored their water not in reservoirs down in the desert but in mountains up amid the clouds.

Although the myth of rugged individualism runs deeper and stronger in the Intermountain West than almost anywhere else in the United States, the reality of life here is far different. The land is so rugged and bare that cooperation is the key to survival. Like the native pueblos and Spanish villages farther east, life in these early Mormon towns was communal, bound together by both faith and water. They surrounded their small towns with irrigated gardens and fields. While families carefully tended their crops, the high plateaus behind them were communal rangeland and woodland, providing grazing for cattle and sheep and trees for firewood and timber. Families often pooled their animals together in town herds. Irrigation ditches were built and maintained by the group. Stores of food were shared with those in need. While pioneers who settled the fertile Great Plains to the east were repeatedly driven from their solitary farms and ranches by dust storms and heavy debts, the Mormons gathered together and learned to thrive in the desert. In time they would push eastward as well, into the red-rock deserts of the Colorado Plateau, building their small towns along rivers and streams amid the canyons—small green oases in a world of solid rock.

Their growing numbers and reach would eventually bring them into conflict with the native peoples around them: the Ute and Paiute of the high plateaus and western desert at first, and later with groups like the Navajo and Apache. While other pioneers often saw the Indians as savages, Mormons believed them to be members of one of the Lost Tribes of Israel, chosen people like themselves. Like the Spaniards, they too sent missionaries to work among them, but conflicts were inevitable and occasionally bloody. There was no surplus or excess to be shared in this new promised land. More land and water for the Mormons meant less for the Indians.

The Mormons, as Wallace Stegner argues in his book *The Gathering of Zion*, would play a leading role in the settlement of the West, opening up not just Utah but a pathway to it as well. Traveling through Iowa, Nebraska, and Wyoming, they made the first roads and built the first bridges. "They were the most systematic, organized, disciplined, and successful pioneers in our history; and their advantage over the random individualists who preceded them and followed them came directly from their 'un-American' social and religious organization." While the Mormons had fled to the desert to escape the outside world, their isolation would not last long. Others would soon follow the routes laid out by Mormon pioneers and scouts, passing through Utah on their way to California and Oregon.

While the Mormons dreamed of building God's kingdom in the desert, those who followed them were often driven by more earthly visions of wealth and power. Once free from the empires of England, France, and Spain, the newly arrived residents of the newly created United States would set about building an empire of their own—a nation that stretched uninterrupted from the Atlantic to the Pacific. Eventually they came to see it as not only economically sound but spiritually sound as well. It was, politicians and journalists of the early 1800s clamored, "Manifest Destiny": The

Americans were a chosen people, blessed with democratic institutions and ordained by God to create a model society in the wilderness. The fact that this wilderness had been inhabited by a constellation of native peoples for thousands of years was merely an uncomfortable fact that could be easily overlooked. While native people like the Navajo and Pueblo believed that they had been created by the gods and present here since the beginning of time, these newcomers often saw their own claim to the land as no less deep; they would bring with them their own stories of migration and struggle, their own gods and myths.

After the close of the Civil War the rush west would begin in earnest. Rumors of silver and gold would bring miners from California to places like Park City and Telluride around the edges of the Colorado Plateau. Booms in cattle and sheep would soon follow as ranchers drove their flocks and herds into the virgin grasslands of the mountains, deserts, and high plateaus. Railroads would connect the Atlantic with the Pacific—one running along the northern edge of the Colorado Plateau above the Uinta Mountains, another running south through Flagstaff and Albuquerque—providing a link with eastern markets and a steady stream of new arrivals. In less than fifty years the plateau country's native peoples would be herded off onto reservations. In little more than a hundred years, its major rivers would be blocked by dams, its deep canyons filled with water. The face of the Colorado Plateau and the West would be changed forever.

E I G H T E E N

Laccoliths and Diatremes

A PARALLEL SET of tracks leads off through the desert across a carpet of broken boulders and stones—the remains of an all but abandoned road. Heading toward the base of Mount Hillers in the Henry Mountains, it crosses and recrosses the deep gully of a dry stream before coming to a dead end at a steep slope of impassable scree—a loose pile of sharp and angular rocks. It flows out of the side of the mountain like a river of broken tiles. Parking the truck, we put on our packs and begin walking, following the loose slope of the scree into a narrow, canyonlike ravine that climbs up the side of the mountain. Walls of rock rise up to either side. They are not smooth and solid like the red and white sandstones of the nearby Waterpocket Fold or San Rafael Swell but cracked and flaked—as if the first puff of wind would send them tumbling down on our heads. From a distance they seem pearl gray. Up close they have a crystalline texture like hard-rock candy, their pale surface peppered with small black crystals.

The rocks here are unlike those found almost anywhere else on the Colorado Plateau. They are not sandstones or shales from ancient deserts and seas but diorite, once-molten rocks that rose up from below, pushing the rocks above them up toward the sky to build up a mountain more than two miles high. Erosion has since stripped most of those rocks away, leaving only high peaks of granitelike diorite behind. Here near the edge of the uplift, however, those molten rocks merely pushed the overlying rocks aside as they rose. Alongside the diorite walls of the ravine you can see the familiar red and white layers of Entrada and Navajo sandstone that cover the slickrock desert of Arches and Canyonlands National Parks to the east. They are not flat and horizontal, however, but tilted almost vertical, thrust up into the sky like a trapdoor.

It is the middle of June, and I am traveling in the Henry Mountains of central Utah with Marie Jackson, a geologist with the U.S. Geological Survey in Flagstaff, Arizona. A chain of five almost solitary peaks, they run through the desert here for more than forty miles: Mount Ellen, Mount Pennell, Mount Hillers, Mount Holmes, and Mount Ellsworth, gradually decreasing in size as one heads south. The tallest is more than two miles high, its high peak covered with forests of aspen, pine, and fir and alpine meadows of flowers and grass. Here to the south, however, the peaks are dry and rocky, covered only by thin, spare stands of pinyon and juniper. High up on their flanks the sites of springs and seeps are marked by small green patches of aspen, visible from several miles away.

Jackson's specialty is the interaction between volcanic and sedimentary rocks. Over the past few years her work has taken her to volcanoes in the South Pacific, Hawaii, and northern California. She began her work right here, however, studying the complex geology of the Henry Mountains for her Ph.D. Filling her summers with fieldwork, she spent the better part of four years mapping and studying the rocks. Today she is back looking over familiar ground, a homecoming of sorts, showing me around the mountains while her husband, Paul Delaney, a fellow geologist at the USGS, watches over their two small children in the campground at the base of the mountain where they first met.

After years of experience working outdoors on shadeless bare rock at high altitude, Jackson is dressed for the sun—not to lie in it but work in it. She is wearing not shorts but khaki-colored pants, and her long-sleeved white shirt is buttoned all the way down to her wrists. A wide-brimmed cloth hat covers her head while a bandana covers the back of her neck. Since ten o'clock this morning the temperature has hovered near a hundred degrees, but Jackson is merely surprised that it is not hotter. "There's no place like the Henry Mountains. They're unforgiving," she says. "But once you get to know them they can be your best friend."

The climb is hard. The loose rocks shift and tumble beneath our feet. In places they sound almost hollow, as if they were made out of glass. A few hundred yards up the slope we stop beneath the shade of a solitary pinyon to catch our breath. Heat waves rise up from the stones, shimmering and dancing in the dry desert air.

Jackson grew up in northern California, where her father, Dale Jackson, was a senior scientist for the U.S. Geological Survey in Menlo Park and a recognized authority on the geology of the Yellowstone National Park area. Noted for both his teaching and research, he taught geology to the astronauts who landed on the moon. Rubbing shoulders with both geologists and astronauts was more than a casual introduction to geology and the earth sciences, but as an undergraduate at the University of California at Santa Cruz, Jackson majored in psychology, not geology. In her senior year, however, she decided she wanted to take a course in geologic mapping. Unable to get into the class at Santa Cruz, she ended up in the University of Wyoming's summer field program taught on the Colorado Plateau.

In graduate school she switched to geology and ended up at Johns Hopkins University in Baltimore, Maryland, working toward a doctorate. Her adviser wanted her to do a "cookbook" dissertation, tackling a straightforward subject related to his own research. Jackson, however, had ideas of her own. Instead of simply going through the motions, she wanted to do something different, studying the relationships between sedimentary and volcanic rocks. For her field area she settled on the Henry Mountains. Few geologists had worked in the area, and her advisers were less than enthusiastic. Finally they gave in but offered little encouragement. "You better do this right," they said. "If this doesn't work you're finished." Asked for advice on planning her fieldwork they told her, "Measure the strain in the rocks."

Sound advice for working in a college laboratory in suburban Baltimore perhaps, but it proved to be of little help out in the field in Utah. While she elected to focus on only the three southernmost peaks, where the rocks were best exposed, her field area still covered well over a hundred square miles. With no one offering advice about how one might go about measuring strain in the rocks for the better part of a mountain range, she set about measuring almost everything else about them: the strike and dip of bedding planes; the trend of joints and faults. Over the course of four summers she logged thousands of measurements. High amid the bare rocks of the peaks, she would work until she was on the verge of passing out from the sun and the heat and then go look for a tree or bush and sit in the shade until she felt stable enough to go back to work. She took field assistants with her as much for safety as help with the work—a collection of friends, relatives, and fellow graduate students. Slopes were steep and the rocks were often loose and unstable. Rockfalls and slides were commonplace. Wanting to make her work complete, she climbed out onto precarious ledges and spurs of rock to take measurements and collect samples. Her first field assistant, an undergraduate from back east, was often so petrified that she stayed several hundred feet below, lying in the sun and working on her tan, calling up from time to time to ask, "Marie, are you all right?"

When no one was available, she worked alone for days or weeks at a time. Although the area was laced with old mining roads, they were often so rough that she found it easier to walk: getting up at four or five in the morning before the sun to head out into the field, only to struggle back into camp around sunset, eat a quick meal, and collapse.

Water was a constant problem. Although well over a mile and a half high, the southern reach of the mountains was still bare and dry. Local ranchers running their herds in the area worried about a young woman working alone in the desert and helped out whenever they could. One had a heavy four-wheel-drive truck with a water tank in back. Whenever she was working in a remote area, he would drop Jackson and her assistants off at a remote base camp with enough water for several days—setting a date and time to return and pick them up. Down at Bullfrog Marina on Lake Powell to do laundry and pick up supplies, she spotted a helicopter from a Salt Lake City television station and talked them into carrying water tanks and a cache of food to the top of one of the peaks where she was hoping to do some work.

Insects posed problems as well. One summer she and an assistant were attacked by clouds of biting gnats. Swollen by hundreds of bites, they became so sick they were unable to hold down food or water. Another year they decided to come out in early spring to try and avoid the summer heat, only to get caught in a snowstorm on top of the mountains. Lost in a whiteout, they finally managed to find their way back down and then drove down to Bullfrog Marina to let people know they had made it down safely—only to find out that no one had noticed they were missing. When she began collecting paleomagnetic samples from the dikes and sills of igneous rocks that had shot through the mountains, a friend in graduate school at Stanford came out to lend a hand. They carried not only the chainsaw-sized rock drills they needed to collect the cores but the water they needed to keep the drill bits lubricated as well, climbing the peaks cross-country with fifty- and sixty-pound packs. Before joining Jackson in the Henrys, her friend had been climbing and trekking in the Himalayas. After a few uneasy nights of sleep amid the boulders and loose rocks of the steep slopes, he announced that he had become convinced that there wasn't a square inch of flat ground in the entire mountain range. The Henrys, he said, were the roughest mountains he had ever seen.

Located in the midst of a barren desert and surrounded by deep canyons and high plateaus, the Henry Mountains were the last mountain range to be added to the map of the continental United States. Floating down the Colorado River in 1869, John Wesley Powell spied the hidden range of peaks from the edge of Glen Canyon. The reports of scouts and explorers had made no mention of them. Powell called them simply the Unknown Mountains. Three years later while triangulating north from Kanab, Utah, and creating the first maps of the canyon country that straddled the Utah-Arizona border, Powell's brother-in-law, Almon Thompson, would find the Henry Mountains again and put them firmly on the map—naming them for Joseph Henry, an early head of the Smithsonian Institution who had been a staunch supporter of Powell's travels and research. Their structure and the forces that had created them, however, were still enigmatic, and in 1874 Powell, now head of a federally funded survey of the Colorado Plateau region, sent Grove

Karl Gilbert off to study the mountains in detail. From Washington, D.C., Gilbert traveled to Salt Lake City by train and then worked his way south along the edge of the high plateaus by mule and pack train. At Salina he turned east and began working his way through the desert toward the Henry Mountains. There were no roads. Instead Gilbert and his group followed Indian trails and game trails, stopping from time to time to explore the canyons and mesas they found around them. The 110-mile trip took more than two weeks. He would spend less than a week in the Henry Mountains, but in the space of a few days he would unravel the details of their complex geology and describe features and forces that few geologists had even seen or contemplated—developing an entirely new theory about the formation of mountain ranges.

Grove Karl Gilbert's career spanned what biographers like Stephen J. Pyne call "the heroic age of geology." Gilbert was born in 1843, ten years after Charles Lyell published the first volume of his landmark book *Principles of Geology*—at that time an unknown field. He died in 1918, ten years before the proceedings of the first international conference on continental drift. Geology at the time was a frontier institution, not the dry, academic world outsiders commonly mistake it to be but topical and controversial, concerned with intensely political and practical matters like finding mineral deposits and irrigable land and creating maps that would define the borders of both states and nations. Prior to the 1850s detailed geographic knowledge was limited to the world's coastlines. Within the continents, only Europe's interior was known with any precision. Everything else was an open book—at least as far as science was concerned. In the United States geology became associated as much with exploration as science as geologists like Powell, Clarence Dutton, Louis Agassiz, and others traveled through the Grand Canyon, Yellowstone, the High Sierra, southeast Alaska, and Hawaii, studying and describing what they saw—measuring the land and the world around them.

Gilbert would make his own mark as an explorer as well, even though his health had been so frail as a child that he had been unable to attend school regularly. Raised in upstate New York, he learned to handle boats in nearby rivers and lakes—a skill that would serve him well on the Colorado Plateau. In college he studied rhetoric, logic, French, German, Greek, and Latin. Science courses consisted of geology, biology, and physics—and the last, more than anything else, would shape his perception of the earth.

Ahead of his time, he thought of the earth as a giant machine. While other geologists of his day studied fossils and tried to determine the earth's age or trace the path of evolution, Gilbert tried to understand what made it work, what forces had given shape to its mountains, lakes, and valleys. In some ways he would have felt more at home in today's world of geology than his own. His obsession with movement and motion and changing forces would, in a sense, anticipate plate tectonics by several decades. While others working around the Great Lakes interested themselves in fossils, Gilbert began studying the contours of ancient shorelines and discovered that the land beneath the lakes was actually rising. Freed from the weight of the overlying ice that had carved them and covered them during the Ice Ages, the ground beneath them was actually rebounding—rising up into the air the way the hull of a ship in port rises slowly out of the water as its cargo is unloaded and transferred to shore. While others traveling through the Grand Canyon marveled at the fossils and built-up layers of rock revealed in the canyon walls, Gilbert thought about ways of using the canyon to calculate the earth's weight. His ideas about the origins of the Henry Mountains were no less mechanical.

The mountains, he quickly saw, were not created by a volcanic eruption but a volcanic intrusion—a column or stock of molten rock that had risen up from below, working its way into the sedimentary rocks above to create a broad dome or swell that caused the ground above to inflate like a balloon. He called them bubble mountains and coined the scientific term laccolites, which would later be altered to laccoliths. Erosion, he theorized, had then carved these broad domes of rock into the sharp-sided peaks seen today. At one time, he speculated, there may have even been laccoliths in the air—ancient mountains that had been completely eroded away. His work in the Henry Mountains would establish his reputation as one of the world's leading geologists. He would later do landmark work on the history of Lake Bonneville, the prehistoric lake that covered much of western Utah during the Ice Ages and a classic study of California's San Andreas fault after the 1906 San Francisco earthquake. An accurate observer with an insightful mind, his notes and papers are still valued by geologists today, nearly a century later.

While adding new details, Jackson's own work would confirm Gilbert's theories of the origins of the Henry Mountains. Each of the three southernmost mountains—Mount Holmes, Mount Ellsworth, and Mount Hillers—Jackson found, were like a snapshot in time. Step by step, they seemed to show how the magmas rose up from below to create the mountains above. Not only are the mountains successively higher as one heads north, but the rocks are tilted at successively steeper and steeper angles as well. Rocks along the flanks of Mount Holmes had a shallow dip of just twenty degrees, while those at Mount Ellsworth dip at fifty to fifty-five degrees. Farther north, on the flanks of Mount Hillers, the rocks are nearly vertical, tilted at an angle of seventy-five to eighty-five degrees. Seen in sequence, Jackson said, each one shows how the rocks were progressively uplifted by the molten rocks intruding from below. On top, however, the rocks are nearly flat, as if the intrusions left the overlying sandstones and shales of the desert here intact, draping across them like a blanket or sheet thrown over the top of a table or chair.

While each of the mountains are different, she added, the interesting thing is that their bases are roughly the same size, between ten and fourteen kilometers in diameter. Rather than a large central stock that pushed its way to the surface, Jackson's work suggests that the mountains were uplifted by a series of smaller intrusions— sheets and fingers of molten rock that worked their way into overlying sandstones and shales above; some known as dikes moving vertically, others known as sills moving horizontally; spreading not upward but from side to side. Fed from a larger body of magma below, these intrusive fingers of molten rock seem to have risen up vertically in dikes until their density was equaled or offset by the weight of the rocks above and then began to spread out horizontally as sills in a disklike fashion. Over time successive dikes and intrusions left these sills scattered throughout the rocks like a stack of plates. In time new surges of magma began pushing these disklike piles of sills upward like a piston, lifting the rocks above them. Peaks like Mount Hillers in the middle of the Henry Mountains were driven upward by more than a mile and a half.

Once the these intrusions began to rise, they tilted the rocks above them like hinges or a pair of swinging doors—not only the overlying sandstones and shales but the overlying stack of sills as well. You can see the outlines of these intrusions in the patterns of faults that surround the edge of the mountains. The faults here do not run in straight lines but are almost perfectly circular—running around the edge of the mountain like a series of concentric circles.

All of this seems to have happened fairly quickly. Paleomagnetic data suggests that the dikes and sills that shot through the rocks were already rotating as they cooled. The mountains here may have taken shape in the space of only a few thousand years. Dating suggests that these intrusions are somewhere between 25 and 35 million years old, roughly the same age as the Rio Grande Rift to the east and the Marysvale Volcanic Field to the west—part of the surge of volcanism that surrounded the edges of the Colorado Plateau with lava and explosions of molten rock. Other laccolithic peaks like the La Sals and the Abajos nearby appeared at roughly the same time. Together with other solitary peaks like Navajo Mountain near Lake Powell and Sleeping Ute Mountain near the Four Corners at the center of the Colorado Plateau, they were part of the region's muted response to the surge of volcanism that swept through the Intermountain West after the close of the Laramide orogeny that shaped the Rockies.

While they never erupted, strangely enough the rate of inflow of magma into the Henry Mountains is nearly identical to that of volcanoes like Mauna Loa on the island of Hawaii several thousand miles away in the Pacific. That incidental similarity is intriguing. The Hawaiian Islands took shape as the moving Pacific plate drifted over a hot spot on the earth's surface. As the plate passed by, streams of magma from below periodically poked through to the surface like the needle on a sewing machine, leaving the islands and their high volcanoes behind. Some geologists and geophysicists have begun to believe that the Colorado Plateau and the Henry Mountains within them may be sitting on top of a similar hot spot as well: a rising plume of magma from the earth's molten interior that is pushing this once stable world up into the sky.

Richard Blank is a geophysicist with the U.S. Geological Survey in Golden, Colorado. He is not, by his own admission, "an old Colorado Plateau hand," but when some colleagues asked him to take a look at the large-scale geophysical patterns in the region, he found some intriguing things. While the well-exposed rocks of the Colorado Plateau offer a wealth of geologic detail, its features and the forces that tie it all together are harder to understand. Blank's work focused not on samples of fossils and stones but on patterns of gravity and magnetic anomalies, data not gathered by months of detailed fieldwork on the ground but in a few hours or days by airplanes flying several thousand feet in the air trailing sensors behind them. But while the data can be quickly collected, processing and analyzing it is another matter, a task for computers and careful manipulations that can take months or even years. Like a satellite photo from space, however, these studies reveal variations in gravity and magnetics that offer a new glimpse of the earth's large-scale features.

On the Colorado Plateau what struck Blank immediately was the way its major features—uplifts, laccoliths, and faults—lined up with major changes in the region's gravity and magnetic fields, as if their broad details had been copied from one another and were almost interchangeable. In addition to marking out its major features, they seemed to wrap around it as well—defining the plateau's edges. The correlations were unexpected. The anomalies Blank was interested in are typically due to changes seen not so much in the earth's surface but several thousand feet beneath it.

Patterns of magnetic anomalies on the earth's surface are generally due to changes or irregularities in the rocks that make up its crust. The earth's metallic core has given it a magnetic field, the force that enables compasses to point north. That field, however, is unstable. For reasons that no one completely understands, every few hundred thousand to every few million years its polarity seems to flip—north becomes south and south becomes north. While that flip has little or no impact on rocks like sandstones or shales, in molten layers of rock like lavas or basalts, the particles and molecules inside the rock line themselves up in orientation to the earth's prevailing magnetic field like tiny compass needles. When the rock cools this orientation becomes locked in place. Not only does this "lock" record the declination or longitude—the direction of north at the time when the rocks cooled—but its inclination, or latitude, as well, its position with respect to north and south. While the forces associated with the earth's magnetic field are oriented nearly horizontally near the equator, at the poles they are almost vertical, curving back into the earth.

Geologists have used this ability of lavas and basalts to record their point of origin to track the movements of continents and growth of ocean basins. At the same time these paleomagnetic patterns can also create magnetic anomalies on the earth's surface. Where this locked-in magnetism is in line with the orientation of the earth's current magnetic field, its effects are additive—creating an area of anomalously high magnetism. Where it is out of phase or out of line with the earth's current field, it creates a band of anomalously low magnetism. Here in the continent's relatively stable interior, these anomalies create a way of tracking the reach of regional features like the Rio Grande Rift and the volcanic fields that circle the Colorado Plateau—not just those lavas or basalts visible at the surface but the large pools or chambers of them beneath it as well. Plotting the region's magnetic anomalies onto maps, you can see the spread of the San Juan Volcanic Field in the San Juan Mountains and the Rio Grande Rift quite clearly. In fact, the two do not appear to be separate but part of a broad belt or welt that wraps around the eastern edge of the plateau, curving through northwestern Colorado all the way to the Uinta Mountains—something not suggested by ordinary geologic maps or mere observations of the surface. Other volcanic fields—the Marysvale Volcanic Field in southwestern Utah and those of the Mogollon Rim—create plainly visible magnetic anomalies as well.

Gravity anomalies, Blank explained, reveal other regional details also not readily visible at the surface. Like the earth's magnetic field, the earth's gravity field is not uniform but often varies slightly from place to place. The reason has to do with the nature of the force itself. Gravity varies with mass. The moon, for example, has a mass that is only a fraction of that of the earth. In fact it is so much less that objects on the moon weigh only one-sixth as much as they do on earth: An astronaut who weighs 180 pounds on earth weighs only 30 pounds on the moon. (In space, which has no mass at all, objects—including astronauts—have no weight at all.) Differences in gravity appear not only between planets but within them as well. Large variations in the earth's crust—areas of heavier and lighter rocks—create gravity anomalies on the earth's surface as well. On the Colorado Plateau, Blank explained, deep faults and uplifts appear as distinctly on maps of gravity anomalies as they do on topographic maps of the region. Prominent uplifts such as the Zuni Upwarp that underlies the Zuni Mountains in western New Mexico or the Defiance Uplift that underlies the Defiance Plateau on the Navajo reservation in northwestern Arizona have brought dense, deep rocks from the continent's craton to the surface, creating areas of unusually high gravity. Low-gravity anomalies, in contrast, mark the region's laccoliths and volcanoes: The hot rocks beneath them are lighter than the rocks around them, creating a correspondingly lower or lighter field of gravity. Here too, however, hidden features emerge. Not only do gravity anomaly maps show the location of the region's laccoliths and volcanoes, Blank said, but they reveal others hidden below—bodies of magma and lava still buried beneath the surface, waiting for erosion or perhaps another surge of activity to bring them to the surface. While Grove Karl Gilbert suggested that there may have once been laccoliths in the air, it appears there are others still hidden below, waiting to make their arrival.

That notion of a laccolith rising to the surface is useful for contemplating the recent geologic history of the Colorado Plateau as well—the forces that began driving it up toward the sky. "The plateau sits on top of a regional topographic high," Blank said. That topographic high, he explained, stretches all the way from Mexico to Canada, its axis running almost right through the center of the Colorado Plateau as well as the Yellowstone area farther north with its geysers and hot springs. At first logic suggests that the highest ground in the West runs through the Rockies in Colorado with its spread of thirteen- and fourteen-thousand-foot-high peaks. When the elevations of its low valleys and parks are factored in, however, the *average* elevation of the Rockies, Blank explained, is actually lower than the *average* of the neighboring Colorado Plateau with its flat mesas and broad desert valleys.

Not only is the area high, Blank added, it is also less dense, a feature that shows up quite clearly on maps of regional gravity anomalies. The Colorado Plateau sits right in the center of this regional high like a knot—the highest and lightest ground of all. The whole region, in a sense, seems to be rising up like a vast flat-topped laccolith covering hundreds of square miles, floating up like a balloon. As it rises, Blank said, its edges seem to be falling apart, collapsing and extending outward in all directions. "I think the Colorado Plateau may have once covered a much larger area," Blank said. "But it's been broken up. The central part, what we call the Colorado Plateau today, has not yet been broken up. It's still coherent. But there's nothing to say that it won't fall apart in the future. That's still in the cards."

There is no mystery as to why the region is less dense, Blank said. It's less dense because it's hot. "The sixty-four-dollar question, however, is what causes it to be hot," he said. Like the reasons for the Colorado Plateau's relative isolation and calm from the violent geologic forces at work in the regions around it, there is no clear-cut opinion among geologists as to what forces are actually responsible for warming the ground beneath it. Some favor continued subduction along the Pacific Coast to the west, others plumes of hot rock within the mantle itself. "My own personal bias lies in the interpretations which favor a mantle plume. People say that's just begging the question because you can't prove a plume. And they're right," Blank said. Bit by bit, however, geologists are beginning to better understand what lies beneath the earth's crust. The mantle is a mysterious place. But while the Colorado Plateau had been creased before by plate movements along the earth's surface, the forces that would finally lift it up into the sky seem rooted in the movements of the swirling hot rocks of the mantle below.

The earth's mantle lies more than twenty miles beneath the surface, a subterranean sea of molten rock enclosing a solid magnetic core. A fluid base for the mobile continents drifting above, the molten rocks within it have a temperature of more than fourteen hundred degrees centigrade. Here on the Colorado Plateau you can see bits and pieces of the mantle right out on the earth's surfaces, not in lava fields or laccoliths but diatremes—holes and pipes left by blasts of superheated steam and loose rock that shot through to the earth's surface. Like the high, solitary peaks of laccoliths, you can find them scattered here and there around the Colorado Plateau: at Shiprock in New Mexico, the Hopi Buttes in Arizona, and in southern Utah near Monument Valley and the Comb Ridge Anticline. The oldest diatremes first appeared between 25 to 30 million years ago, part of the Colorado Plateau's muted response to the waves of volcanism that swept through the Intermountain West; roughly the same age as the laccoliths that dimpled the center of the Colorado Plateau with solitary peaks.

In places like South Africa diatremes left behind diamonds that took shape under extreme high pressure and heat. Here they blew peridotite—crystalline rocks from the earth's mantle—up to the surface through more than a dozen miles of rock. They appear not as sharp-sided blocks or chunks but rounded pebbles and stones of exotic rock. Tumbling and rolling along the sides of a deep channel or pipe as they blew to the surface, they were worn almost smooth. While jets of superheated steam and bits of the mantle blew up from below, loose blocks of sandstone tumbled back down into these pipes like loose rock in a mine shaft, falling more than a mile.

Seen from above they look almost out of place. From the air those like the Mules Ear diatreme near the San Juan River in southern Utah look like nothing more than a small steep hill of blue-green rock—as if a giant haystack rock from the California coast had been dropped in the midst of the desert. On the coast those giant stones are made up of twisted rocks from the subduction zone. Subduction carries not only pieces of the seafloor down into the earth beneath the edge of the continent but seawater as well. Here it looks almost like that mixture of watery rock has blown through to the surface nearly a thousand miles from shore, poking through the red, slickrock deserts of the Colorado Plateau with an incongrous pillar of green. That similarity to rocks found near the subduction zone is due to more than just coincidence.

Some 30 million years ago, subduction along the California coast came to a close as the western edge of North America ran over the East Pacific Rise. The angular collision split the rise in two—one heading north toward Washington and Oregon, another heading south to Mexico where it would eventually separate Baja California from the Mexican mainland to create the Gulf of California. In between them the San Andreas Fault appeared, opening like a zipper between them. From Mexico to northern California the edge of the continent was no longer bordered by a plate of oceanic crust that was diving beneath it but by a plate that was moving northward alongside it.

Farther inland the end of subduction seems to have had its own effects as well. As the trailing edge of the subducting plate passed by, it left a gap or window behind it that was filled by the mantle rising up from below, lifting up not only the Sierra Nevada and the neighboring Basin and Range but the Colorado Plateau as well—or so many geologists believe. While 20 million years of shallow subduction crumpled the earth's surface to lay the base for the modern Rocky Mountains above, it sanded the rocks below away, thinning the earth's crust and leaving the hot rocks of the mantle several miles closer to the surface, their heat causing the earth's crust above them to rise.

The idea of a gap or slab window playing a role in the final rise of the Colorado Plateau is an elegant and intriguing theory, but one whose existence and effects are hard to prove. While scientists like Richard Blank with the U.S. Geological Survey have spent time studying the Colorado Plateau's magnetic and gravity anomalies for clues about its structure and origin, others like Jon Spencer with the Arizona Geological Survey have tried to understand the forces responsible for its uplift by developing complex computer models to predict the effects of a slab window on the Colorado Plateau. The project took three years. The first was spent simply understanding the processes at work, the properties of the earth's mantle and crust and calculating the regional uplift; the second writing the programs that would weave all these ideas together; and the third simply running them through the computer to come up with results that could be published in a paper.

Manipulating the data from digitized maps, he found that the plateau had been uplifted by an average of anywhere from thirty-five hundred to nearly forty-five hundred feet. His most optimistic models of uplift, however, could account for only half of that uplift—some twenty-four hundred feet—factoring in such things as not only the heat in the mantle but additions to the earth's crust due to volcanic eruptions and the alteration of rocks below by water. While the rest of that uplift could be caused by a plume of unusually hot rock in the mantle, the truth of the matter is anyone's guess. "We just don't know," Spencer said. "It's a damn mystery and that's that."

It's tempting to do more modeling and research to solve this problem, but Spencer believes that further work will have to wait until more is known about the actual physical properties of the mantle. "You can make all kinds of speculations," he said. "And then somebody comes along who looks at real rocks and tells you that the correct number is twenty not five to start with, and you suddenly find that you're out on a long limb and he's busy sawing at it."

Whatever its causes, the uplift of the Colorado Plateau would mark a sharp break with the past, an end to several hundred million years of almost uninterrupted deposition and collection. As the Colorado Plateau began to rise up toward the sky, its rivers and streams would begin cutting down through the rock, carving out canyons, mesas, and buttes—to create a surreal world of solid rock.

N I N E T E E N

Mountain Forests

IT IS A HALF hour before sunrise and the air is quiet and still. The day before wind gusts of thirty to forty miles per hour had made it almost impossible to stand. Behind my makeshift campsite above ten thousand feet on top of the Wasatch Plateau, a line of Engelmann spruce had swayed in the wind like reeds. This morning there is no sound at all on top of the plateau, and the sky overhead is indigo blue. To the east it has already begun to lighten and glow, the edge of the horizon marked by glowing bands of red and pink. In the faint early-morning light I can make out the rectangular outlines of distant mesas and buttes in the desert below, outlined by the brightening sky behind them. To the west the sky is still jet black. Beneath it the high, jagged lines of mountains mark the edge of the sky, blue-gray shapes in the twilight, running west as far as the eye can see like sea swell, waves of rock rolling through the desert.

As the sun begins to rise, the plateau comes to life. Amid the bare, rocky fields and tundralike meadows that cover the top of the plateau, horned larks scurry across the ground while mountain blue-birds appear amid the spirelike branches of the spruce and sing. Out amid the boulder fields and scree slopes that lead down into the forests below, pika and ground squirrels scurry from rock to rock, calling to one another with high-pitched whistles or squeaks. The week before in the desert the temperature had climbed to 105 degrees. Here on top of the Wasatch Plateau it barely climbs above 60, falling to well below freezing at night. Although it is late June, patches of snow still cover the ground. Some have already begun to melt, leaving behind small lakes and pools. Others are still white and deep, stretching across the high line of ridges in deep snowbanks that will last until well into July. As I load up a small daypack for a morning walk along the crest of the plateau, I find that my water bottles are rimmed with ice.

The western edge of the Wasatch Plateau rises up in a sheer high wall, its flat top several thousand feet above the valley below. To the east it slopes gently down into the dry, rocky world of the Colorado Plateau. In spite of the surrounding deserts, snow is an almost constant part of the landscape here, gone for no more than two or three months of the year. Aside from a few thin lines of twisted, windblown trees, the ground here is as open and bare as arctic tundra, a world of grass and flowers that springs to life for only a few weeks each year. Down below, between the desert and the snow, the forests and meadows are already in full bloom. The new leaves of the aspen are a bright golden green, set off by the dark, deep green of spruce and fir. Alpine meadows of rich, green grass run through the forest like ribbons. Traveling across the top of the plateau, I find myself walking through fields of flowers: yellow glacier lilies and white marsh marigolds, wet, green meadows fringed by fields of snow. Looking up from the flowers I can see the bare, red rocks of the desert below.

The high peaks of the western United States squeeze scarce moisture out of the sky. The wall-like rise of mountains and plateaus drives prevailing westerly winds upward. The air cools as it rises, reducing its ability to hold moisture, bringing rain and snow to the ground below. Like a windbreak blocking the wind, these high peaks block the rain, creating a "rain shadow" along their leeward sides—a corresponding zone of low rainfall. The deserts of Nevada and western Utah are the result of the rain shadow cast by the Sierra Nevada farther west, while the Rockies cast their own rain shadow over the Great Plains as well. By the time prevailing westerly winds reach the western edge of the Colorado Plateau they have drifted across several hundred miles of desert and dozens of mountain ranges. As the hot, dry wind blows across the tops of the plateaus, clouds of rain and snow appear almost like magic. The ground is not only wetter here but cooler as well. Average temperatures drop three degrees for every thousand feet of elevation, and the crest of the plateau here is more than six thousand feet higher than the desert floor below. While the bare rocks of the desert burn in the midsummer sun, the tops of the high plateaus are cool and green.

The effects of changing altitude on precipitation and temperature were first noted by the German explorer Alexander von Humboldt, who traveled throughout Central and South America in the mid-1800s. While the coastlines were often covered by rain forests, the tops of the region's high peaks were covered with snow and ice and alpine tundra. In between were a host of steadily changing forests and environments.

The effects of changing altitude on the Colorado Plateau, with its deep canyons and high plateaus and peaks, is no less striking— ranging from cacti and dry scrub on the valley and canyon floors to alpine mosses and flowers on top of the region's high peaks and plateaus. The phenomenon was first described in the 1890s by the naturalist C. Hart Merriman. Working in the San Francisco Peaks area north of Flagstaff, Arizona, he divided the steady change from desert to forest he encountered while climbing the peaks into a series of "life zones," correlating them to changes one might see traveling northward across the continent. The grasslands and scrub-lands of the desert floor became known as the Lower Sonoran, the pinyon-juniper forests above them the Upper Sonoran—their names reflecting their resemblance to the Sonoran deserts of northern Mexico and southern Arizona. Above these scrublands and wood-lands were forests of ponderosa pine and Gambel oak, a border between the desert below and the mountain forests above that Merriman called the Transition Zone. Beyond it one seemed to have been magically transported into the woods of northern Canada—a rich, wet forest of aspen, spruce, and fir known as the Canadian Zone and an even higher one where the same tall trees began to shrink in stature and size, stunted and twisted by high wind and heavy snows. Known as the Hudsonian Zone for its resemblance to the stunted forests that border Hudson Bay on the edge of the Arctic, it was marked as well by the appearance of long-lived trees like limber and bristlecone pine. On the top of the peaks there were no trees at all, only the mosses and low grasses of the Arctic or Alpine Zone—a world of snow and tundralike turf like the northernmost reaches of Canada and Alaska above the Arctic Circle.

Mountain forests blanket the Colorado Plateau above seventy-five hundred feet, and in places where the land rises up above eleven thousand feet there are often no trees at all. You can find forests of ponderosa pine on the North Rim of the Grand Canyon in Arizona and on the flanks of Mount Taylor in New Mexico. Forests of aspen, spruce, and fir blanket the flanks of the Aquarius Plateau in Utah and the Uncompahgre Plateau in Colorado. Alpine meadows and patches of tundra are found not just here on the gently rolling crest of the Wasatch Plateau but far to the east as well on the high, steep slopes of the La Sal Mountains overlooking the slickrock deserts of southeast Utah and its tangled maze of canyons.

The boundaries between this succession of ecological zones are often sharp and distinct. The deserts and forests here, as mentioned before, are often as layered as the rocks beneath them. They are not unrelated. Hidden in the shade of alcoves and deep canyons carved into the rocks of the desert below you can still find relict forests of pine and fir like those found amid the mountains and high plateaus, left over from the Ice Ages when these mountain forests reached far down into valleys below. Fifteen thousand years ago the tops of these high peaks and plateaus were bare, covered by permanent fields of snow and icce, their flanks carved by glaciers that slid down their sides and into the broad valleys below. As the climate began to warm, desert plants pushed into the valleys below while trees began to climb up the flanks of the high peaks and plateaus. Today these mountain forests are a counterpoint to the desert below, a source of water and life, a high green roof for the red-rock walls of the desert.

The transition forest that marks the start of this mountain world over much of the plateau is not low and close like the pinyon-juniper woodlands of the desert but high and open. The tall, straight trunks of the ponderosa pine tower above the forest floor, their tops more than eighty or a hundred feet high. The trunks of the oldest trees are four and five feet in diameter, their plates of reddish brown bark divided into squares by a gridwork of furrows and cracks. Its corklike surface seems almost tiled, the thick bark offering protection from the fires that periodically sweep across the forest floor. Its needles are not short and stiff like those of the pinyon below but silky and long, clustered together in bunches of three or five. Cones litter the ground beneath the trees like giant eggs five and six inches long.

Although the forest here typically receives more than twice as much rain as the deserts below, the ground is still dry. Less than twenty-five inches of rain falls here each year. Fire keeps the forest open and parklike, clearing the ground of scrub and the tangled branches of dead, fallen trees. After the spring snow melt it bursts into bloom, a floor of green grass and bright flowers: larkspur, sego lilies, and Indian paintbrush.

Elsewhere this zone of transition between desert and forest is quilted by thickets of scrub and low trees: dense, head-high stands of Gambel oak whose acorns provide food for wild turkey, grouse, and deer. In fall their chestnut brown leaves are set off by the ever-green branches of the pines. Tough shrubs cover the drier ground here as well—tangles of mountain mahogany, blackbrush, and sage—but they fade as the land climbs up into the rain.

While thickets of Gambel oak are found throughout the Rocky Mountains of Colorado and the neighboring Colorado Plateau, aside from a few pockets in southern Nevada, they reach no farther west than the high plateaus of central Utah. In contrast, the ponderosa pine is one of the most widely distributed trees in the West, found all the way from Canada to Mexico wherever rainfall and soils are sufficient. (On the Wasatch Plateau soil chemistry inhibits the growth of the pines, and the Transition Zone is dominated by tall stands of Gambel oak.) Named for its heavy wood (*ponderosa* means "ponderous") it is valued as timber and is one of the most heavily logged trees in the West. Here on the Colorado Plateau these transition forests of pine and oak dominate the landscape between seven and eight thousand feet.

After the sporadic world of the pinyon-juniper forest below, life here seems rich and abundant. Western tanagers and Steller's jays flock to the rich world of the transition forests. While woodpeckers pound the tall trunks of the ponderosa pine, flocks of wild turkeys run through neighboring stands of oak and brush.

As the land climbs higher, the forests of pine and oak give way to dark groves of spruce and fir and bright forests of aspen. Aspen are perhaps the most widespread trees in North America, found all the way from the mountains of northern Mexico to the Brooks Range in Alaska on the edge of the Arctic. Here amid the mountains and high plateaus of the Southwest they can be found anywhere from seventy-five hundred feet to ten thousand feet in elevation, where the average rainfall is typically more than twenty-five inches per year. Deciduous trees with straight white trunks, their leaves are set off by the dark needles of the spruce, pine, and fir that surround them—bright green in summer and gold in the fall. On a clear day when gusts of wind play along the slopes, their leaves seem to shimmer and dance in the sun.

Unlike the evergreen trees that surround them, aspen thrive in disturbed areas: the sites of burns and avalanches. Although they produce thousands of seeds, they are seldom fruitful. Instead, they most often reproduce by cloning, sending shoots and roots that give rise to new trees, like grass in a city park.

At times walking through the aspen forest feels like walking through a hall of mirrors. Their straight trunks reach up sixty, seventy, and eighty feet like poles toward the spreading canopy of leaves overhead. Their white bark looks almost like paper. It peels away from the trunk in thin sheets. Here and there you can see messages carved in the trunks of the trees by passing sheepherders and hunters: a collection of dates and initials, even whole poems in Spanish, signposts amid the forest faded with time into black furrowed scars. Beyond them the straight white trunks seems to stretch off in every direction, an endless forest of identical trees. In fact, the trees here not only look alike, they quite often are alike, genetically identical copies of one another, sprung from a common center or root. These clusters of cloned and copied trees are known as *genets*, while the individual trees themselves are *ramets*. The forests here are not simply a collection of trees but a single living organism covering several acres or even several hundred. The forest and the trees here are all tied up together.

Travel through the mountains and high plateaus in the early spring, said Walt Mueggler, a scientist with the U.S. Forest Service's Forestry Sciences Laboratory in Logan, Utah, who has spent several decades studying aspen populations in the Intermountain West, and you can see how the trees within a particular genet or group leaf out at the same time. The synchronization is due not so much to temperature or rainfall but the genetic patterns and codes locked up in the trees themselves. Other neighboring stands of trees can remain bare for days or weeks, as each cluster of trees follows its particular schedule. The synchronization extends into the fall as well when the trees change color, their leaves turning from green to bright gold. Look closely, however, and you will notice subtle differences in color between the different genets of trees on the side of a mountain or plateau. In places Mueggler has even found stands of aspen in the wild whose leaves turn bright red. Trunk color varies as well between genets: some pure white, others tinged with shades of yellow or green. The details are all identical. The trees have the same color, height, shape, and age.

While individual aspen are seldom more than a hundred years old, biologists believe that the roots they spring from may be more than a thousand years old, derived from seeds that germinated when the climate of the Intermountain West was wetter than today. New trees spring from horizontal root suckers and runners that spread out from the base of the trees a few feet below the surface. Resistant to fire, they can reach some one hundred feet. In places you can actually see how the trees spread: a core of old, tall trees surrounded by a thicket of saplings.

After an avalanche or fire, new shoots spring up quickly from the runners and suckers below, creating a dense stand of saplings. After a decade or two the saplings begin to thin out. Those that survive begin rising, reaching sixty, seventy, and eighty feet up into the sky. To curb competition, they release chemicals known as auxins into the soil to inhibit the growth of new aspen from runners that still lie beneath the forest floor. Instead, seedlings of spruce and fir appear beneath the trees. Shade-tolerant trees, they grow slowly among the aspen, rising up year by year until they finally reach above them, shading them out to create an evergreen forest of dark, dense trees.

Decades of fire prevention, however, have curbed the frequency of forest fires in the West—and as a result aspen forests have been decimated. While the roots and suckers can survive for as long as fifty years underground after the trees above them are gone, in places the trees have been gone so long that their roots and runners have vanished as well. The patchwork of aspen and evergreens that once covered the mountains and plateaus of the West has slowly given way to a dark solid forest of spruce, pine, and fir—with the aspen among them shrinking year by year. Grazing has played a role in their demise as well: The young shoots are prized by cattle who repeatedly crop them back, making it impossible in some areas for the trees to spread. Fortunately, the problem is recognized. Changing patterns of management and use may, in time, help bring them back.

The forest floor here is rich and green—as varied as that beneath the ponderosa pines below. While grass and sage blanket the ground beneath the trees in lower and drier areas, higher up they are replaced by ferns and blueberries. In winter the aspen forest is clean and open, its straight white trunks rising up through drifts of snow. In summer it is rich and crowded with life. Shaded patches of bluebells and wild geraniums bloom beneath the trees amid a carpet of grass and fern. The rich green world of the forest floor is a mirror for the green canopy of leaves overhead. In spring elk gather in the shade and protection of the trees to calve and raise their young. In summer black bears feed on berries and roots. Birds thrive here as well: red-naped sapsuckers, violet green tree swallows, mountain bluebirds, and dozens of others. Some migrate up and down the flanks of the plateaus and peaks with the seasons, others like hummingbirds and warblers—small, brightly colored songbirds that feed on insects in the tops of the trees—spend their winters in the deserts and rain forests of Mexico and Central America.

While the white-trunked glades and galleries of aspen here are filled with color and light, the forests of spruce and fir that surround them are quiet and dark. Above nine thousand feet they become increasingly prevalent, leaving the aspen to the steep narrow reach of avalanche chutes—clearings cut by the paths of avalanches on the sides of the high peaks and plateaus. The closely packed trees, dominated by stands of Engelmann spruce and subalpine fir, shade the forest floor. Even in midsummer the ground beneath them is damp and cool. Others are here as well: tall stands of Douglas fir and blue spruce. As much as forty inches of rain falls here each year. In winter the ground is covered by as much as five feet of snow. The average year-round temperature here is only a few degrees above freezing.

At first glance the spruce and fir look like nothing more than high, narrow pines. Their spirelike shape is a product of the heavy winter snows as they shed deep piles of snow like a steeply pitched roof. Their needles are not long but short and stiff—tightly packed on their flexible branches like the bristles on a brush. Those of the spruce often feel prickly to the touch, while those of the fir are softer. Cones are different as well: While those of the fir are brown and tend to hang down from the branches, those of the spruce are purplish and stick almost straight up into the air, growing amid the high branches of the trees. Unlike the aspen, which must cover themselves with leaves each spring, the spruce and fir can draw energy from the sun as soon as the air begins to warm. Chemicals within the needles and branches keep them from freezing solid in the subzero temperatures of winter.

In summer the dense carpet of needles that covers the forest floor deadens sound. High up in the trees kinglets, chickadees, and titmice flit from branch to branch, at times the only sign of life. Although the ground is moist here, the climate is harsh. It is only higher up, where these dense forests of trees begin to mingle with the barren alpine world, that it begins to seem truly alive. In midsummer the border between the dark forest below and the bright tundra above draws animals by the thousands. Fed by the sun and the melting snow, this cold, high world is transformed into a rich, green garden, a refuge from the hot, dry land below.

"Here in the West people have very little appreciation of mountains," said Ray Brown, a biologist with the U.S. Forest Service's Forestry Sciences Laboratory in Logan, Utah. "There's no getting around the role mountainous terrain has played in where people have settled," he said. "Water is the key to human life. In the West mountains are what make life possible."

Brown grew up in southern California, where his grandfather worked for Los Angeles County's Metropolitan Water District, the agency that manages and maintains the intricate system of reservoirs and canals that bring the city its water. Traveling with his grandfather to the reservoirs and lakes in the mountains outside town, he grew up understanding the link between mountain and desert that runs through much of the West. There was no need for complex lectures or explanations. "You could go up into the San Gabriel

Mountains and there it was," he said. Pipelines and canals from the reservoirs and lakes led down from the mountains, carrying water to the cities and towns below. "It was all right in front of you."

"People have gotten further and further away from that," Brown said. They shuttle between air-conditioned homes and offices in air-conditioned cars. "They've insulated themselves from the outdoors," he said. "They have no appreciation of how it all came to be."

In the Intermountain West today mountains are all too often viewed as scenery—cheap land for condos or vacation homes, or rangeland and timberland to be auctioned off at bargain-basement prices. It is not just the high reach of the mountains or the deep snows that cover them that are important but literally everything about them. While the population of the West is rapidly growing, few understand how completely their lives are tied to those of the mountains around them. The forests, meadows, and soils that cover their flanks are not just ground cover but hidden storehouses of water. Instead of vanishing in a flash flood or evaporating into thin air, the rain and snow that falls in the mountains filters through them, feeding the springs and streams that water the deserts below.

Brown's research has focused on finding ways to speed the recovery of mine-damaged land. His work has taken him all over the West from uranium mines in the desert to silver and gold mines in the mountains. For the past few years he has often found himself working above timberline—ten, eleven, or twelve thousand feet up—amid the alpine world of grass, bare rock, and tundra that lies above the reach of the trees. The problems he sees are often alarming. As in the deserts below, life in this extreme world is both tenacious and fragile. A few months of abuse can affect the land for centuries—poisoning streams and stripping hillsides of flowers, plants, and soil. As different as it is from the world below, the plants and animals of this alpine world are linked to the dry land below, not just by the water that flows between them but by the way they struggle to survive.

Above ten thousand feet on the Colorado Plateau the forests of spruce and fir begin to slowly decrease in size, shrinking from tall spires that tower seventy and eighty feet above the ground to small trees barely more than head high. Bristlecone pine appear amid the stunted trees. Tough, slow-growing trees, they thrive along this high edge of the forest and are seldom found below ten thousand feet. They are some of the oldest living things in the world. In California bristlecone pines more than four thousand years old have been found amid the high peaks of the Sierras.

Although the ground here near the edge of the trees is high, rainfall is actually five to ten inches less than in the richer forests of aspen, spruce, and fir below. The high peaks and plateaus draw not only rain and snow but wind. While winter storms blanket the peaks with snow, the high winds often strip the slopes clean, leaving the trees exposed to weather while the ground freezes solid beneath them. Higher up flag trees appear with branches growing only on their leeward sides. Pruned by the wind, they look almost like weather vanes. The branches that face the prevailing winds have all been killed by exposure. Higher up others seem to crawl across the ground like woody vines, twisted and stunted by the wind and extreme cold into a carpet of *krummholz*, or bent wood. They huddle in the lee of boulders and draws, rising only a few inches above the ground. Looking at their bent shapes, you can almost see the push of the wind.

The high plateaus of Utah flirt with the edge of this high, bent forest. In places the ground here rises up still higher to reveal an open, rolling alpine world of grass. Walking along the crest of the Wasatch Plateau at nearly eleven thousand feet gives one the feeling of endless space. It is not just the views that extend off in all directions or the open reach of the land but the clarity of the air and light. The air is thin here. There is no haze of moisture or dust. Above ten thousand feet there is two times as much ultraviolet light as there is at sea level and 25 percent more light. The sharpness of the air and light distorts all sense of perspective and scale. Objects several miles away seem like they are only a few minutes away on foot. There are no trees to separate one from the sky or define the edge of the horizon. The world seems closer to the sky.

Snow is a constant part of the landscape here. It arrives as early as late September and often lasts until well into July. Melting drifts of snow leave behind pools of water that are later replaced by beds of flowers. Freezing and thawing fractures boulders and stones, creating vast empty fields of bare, broken rock. Elsewhere, where the high ground is covered with soil, alpine meadows of flowers and grass appear. Other areas are somewhere in between—stony fields or fell-fields where the frost has caused stones from beneath the surface of the soil to heave, creating polygonal patterns of uplifted stones. For the space of a few weeks the land here is rich and green. During the brief days of summer the high ground here bursts into flower, like the desert below after a sudden rain.

That yearly rich explosion of life on the high plateaus attracts a menagerie of animals: elk that graze in the high meadows above the trees and marmots that sun themselves on tops of boulders and stones. Elsewhere pika scurry from place to place, cutting grass and carrying it back to the mouths of their dens to build up small haystacks, food for the long winter ahead. Digging a network of tunnels beneath the snow, they remain active all year. While jackrabbits in the deserts below have oversized ears to help keep them cool, the small ears of the pika help reduce heat loss. For the space of a few weeks the land here is rich and green.

Birds fly up here as well to feed, drawn by flowers, insects, and seeds: flocks of mountain bluebird, horned larks, and rosy finches. Others are concentrated near the timberline, on the border between the trees and the tundra: Clark nutcrackers and Townsend's solitaires. Only the pheasantlike ptarmigan is a year-round resident here, changing its plumage from white to brown with the changing seasons. Standing on the edge of a meadow, black-chinned and calliope hummingbirds streak from flower to flower, gathering nectar and food. They seem incongruous and out of place above ten thousand feet. By winter they will be several thousand miles to the south in the mountains of Mexico.

The plants, of course, are what make this burst of life possible in this extreme world beyond the trees. Fifteen thousand years ago during the Ice Ages the high peaks and plateaus here were covered with glaciers. You can still see their effect in giant boulders and stones that litter the basins and valleys several thousand feet below. In the La Sal Mountains they carved the high peaks into sharp angular shapes. On the flat surface of Thousand Lake Mountain the passing ice left a series of small lakes behind. The land here is still a frontier. Plants and animals have only recently been able to settle it.

The world above timberline is one of the most uniform ecosystems on earth. You can walk across open tundralike landscapes that seem almost identical to those found on top of the high plateaus and peaks of the Colorado Plateau not only in the alpine reaches of the nearby Rockies and the Sierras but also in Lapland, Greenland, Alaska, and Siberia. The plants here are all direct descendants of those that huddled around the edge of the ice that once covered this high ground several thousand years ago.

Like the desert below the ecosystem here is simple and direct. There is no complex layering of plants. Life is deceptively clear and straightforward. Like cryptogamic soils in the desert, lichens cling to the sides of boulders and stones, covering them with bright splotches of color: rust, yellow, lime green, even black. In places they are joined by deep green cushions of moss. They capture wind-blown bits of sand and organic debris, building up thin layers of soil—although soil formation here is incredibly slow, as little as an inch every thousand years, according to Brown.

Snow is a source of both water and warmth above timberline, shielding the plants from wind and bitter cold in the winter. In early summer as the ground begins to clear, they reflect heat and light from the sun, warming neighboring patches of ground. In places they melt, leaving behind a thin layer of ice that turns the ground beneath it into a greenhouse—shielding the emerging plants below from the wind and cold while letting the sun pass through.

When the snow clears everything seems to happen at once. While the growing season at five thousand feet in the valleys below can be as long as two hundred days, here on top of the high plateaus above ten thousand feet it seldom lasts for more than thirty. Frost is possible every night of the year. This growing season is not defined by frost-free days but by the growing range of the plants themselves. While most plants need temperatures of forty to fifty degrees to begin growing, plants above timberline begin growing almost as soon as the temperature climbs above freezing. By the time the daytime temperatures reach fifty degrees, most plants here are already growing at their peak. In midsummer the soil that anchors their roots seldom warms to more than thirty-five degrees. These alpine plants are so well-adapted to this cold, high world, however, that they seldom survive when transplanted to richer and warmer soils, even in the carefully controlled world of a greenhouse. Like plants in the desert, they are carefully attuned to the environment around them.

High altitude is a high-stress environment, and the plants here have fewer stems and leaves than those found at lower altitudes. Flowers, however, are gigantic with respect to the size of their stalks and leaves, a trait that allows them to gather heat and energy quickly. While wind pollinates the grasses here, the flowers often depend on bees and hummingbirds, and their bright colors and patterns are keyed to attracting them. Although the hummingbirds are migrants from Mexico, most of the bees found above timberline live here year-round, their hives hidden underground in burrows. Like the flowers, they are unusually large, a design that allows them to warm up quickly in cold midsummer mornings and take to the air making their rounds of the flowers.

The surfaces of plants are adapted to the cold as well: The stems and leaves of some are covered with fine hairs that provide a protective layer of warmth and help reduce water loss. Although the land here is cool, the constant wind and thin, dry air mean that the evaporation rate, like that of the desert below, is incredibly high. To reach needed water, small flowering plants like spring beauty can have taproots reaching down six feet below the surface. Others like stonecrop rosettes have sticky surfaces that hold on to water.

This surge of life lasts for little more than a month. In a few weeks flowers will wither and die, their stalks blown by the wind. Meadows of green grass will turn bright gold, then brown. Before long the snows will return: a light dusting at first, followed by heavy storms. In winter, the life here moves to warmer ground below. While summer visitors like elk and hummingbirds have all moved off to warmer ground, the grasses and flowers here are all buried beneath the snow, hidden away in the ground in roots and seeds. After ten and a half months of sleep they will awaken again and burst into flower, ready for a month and a half of life. As the snow on top of the high plateaus begins to melt, the water will flow down into the deserts below.

T W E N T Y

Mormons

RICHARD STEVENS cajoles his four-wheel drive around the deep drifts of snow that block the road leading up to the top of Tent Peak, a low hill above the trees that marks the highest reach of the Wasatch Plateau. As we crest the top of the peak the Sanpete Valley spreads out beneath us. Hills as steep as cliffs lead down from the snowfields and alpine meadows that cover the ground here above eleven thousand feet, leading down into forests of aspen and fir. Six thousand feet below the floor of the Sanpete Valley is a desert of grass and sage. Scattered across it is a string of small towns: Manti, Ephraim, Moroni, and Fountain Green. Settled by Mormon pioneers in the 1850s, they lie at the mouths of canyons along the base of the plateau, surrounded by the orderly rectangular shapes of irrigated fields, a quilt of green and brown. On the far side of the valley the San Pitch Mountains rise up to more than eight thousand feet, the start of the Basin and Range, a world of high mountains and desert valleys that reaches all the way to California. Down along the base of the plateau you can see the desert give way to forest—the dry grass and sage fading into forests of pinyon and juniper that lead up into the wetter and greener world above.

Stevens grew up in Ephraim, Utah, at the base of the Wasatch

Plateau. Aside from his time as a Mormon missionary in northern California, a brief stint with the U.S. Forest Service in Salmon, Idaho, and graduate work at Brigham Young University in nearby Provo, Utah, he has spent most of his life right here in the Sanpete Valley. He did his undergraduate work at Snow College, not far from his office in the center of town with the Utah Fish and Game Department. His home sits at the mouth of Ephraim Canyon near the base of a small hill used by the town's first settlers as a lookout during Utah's Black Hawk War to watch for marauding bands of Ute sweeping through the valley or down from the plateau above.

Stevens's area of expertise is wildlife and range management. We had spent the morning driving around the plateau and stopping to look at plantings of grass and trees, talking about everything from winter rangeland for elk and deer to polygamy in the early Mormon Church. While Stevens has spent a great deal of his professional career trying to understand the natural history and ecology of the Wasatch Plateau, his interest in local history is equally deep. In fact the history of the plateau and the town are so closely tied together that they are often difficult to separate. "The Sanpete Valley was settled by people who came from the desert to the protection and isolation of the plateau country," Stevens said. "They built their homes near the canyon mouths on the valley floors. Their lives in the desert were tied to the high plateaus that rose up behind them."

The first Mormon wagon teams reached Utah in July 1847. Fleeing religious persecution in Illinois, they traveled across the Great Plains and the Rocky Mountains to the edge of the desert, driven by visions and prophesies and dreams of building a New Jerusalem.

Founded in 1830 in western New York as the Church of Jesus Christ of Latter-Day Saints, the Mormon Church had an amazing ability to attract both followers and enemies. In less than a decade its ranks of followers grew from six to several thousand while the church itself became a lightning rod for violence and persecution. In 1832, church founder, prophet, and leader Joseph Smith was tarred and feathered in Ohio. A year later Mormon colonists near Independence, Missouri, were driven from their homes by angry mobs. These early events set the tone for the harassment that would plague the church for the next several decades. As attacks from outsiders increased, they organized their own militia, the Sons of Dan, and mounted counter-raids of their own—a continuous circle of beatings, barn burnings, and swapped brutalities that continued for nearly a decade. Revelations of polygamy among church leaders and financial scandals raised the suspicions and resentments of outsiders even further.

In the midst of all these troubles the church thrived. Chased out of Missouri, they settled in central Illinois and founded the town of Nauvoo on the Mississippi River. In less than five years it became the largest city in the state, with a population of more than twenty thousand. Meanwhile converts by the thousands arrived from England and Europe, where church missionaries were hard at work.

Eventually Smith's own heavy-handed efforts to quell dissension within the ranks of his followers would bring trouble with the law along with charges of polygamy. With Mormon and state militias on the verge of open war, a warrant would be issued for Smith's arrest. After fleeing to Iowa, Smith and his brother Hyrum surrendered to authorities, but they would last only a few days in jail. In June 1844 they would be shot to death by a blackface mob in the jail at Carthage, Illinois. The next year in 1845 the state legislature would revoke Nauvoo's charter. Under the terms of an agreement worked out between church leaders and the state governor, the Mormons would leave Illinois the following spring, "as soon as the grass grows and the water flows," heading west toward Utah in search of the promised land.

The early 1800s were a time of intense religious fervor, with sects, orders, and groups of all kinds trying to create an earthly utopia free from the constraints of the past. Joining the Mormons in this quest were New England Transcendentalists, Shakers, Campbellites, and others. Many of those who made the first long trip to Utah had gone on long spiritual journeys of their own before joining the ranks of the Latter-Day Saints—sampling churches and religions the way some might try on clothes or taste wine—looking for the right religion, the right way of life and the right people. While other movements of the day faded, however, the Mormon Church endured, quite possibly because, as Wallace Stegner writes in his book *The Gathering of Zion*, "It had everything other religions had and more: not only total immersion, seizures, the gift of tongues and other aspects of the Holy Ghost, baptism for the remission of sins, and the promise or threat of the imminent Second Coming, but also true apostolic succession and the renewal of the ancient personal communication with God."

Smith's new church preached that other Christian churches had lost their authority through apostasy a few centuries after Christ's Assumption. The religious authority that had been passed down through the apostles, through Peter, James and John, had been passed directly to the head of the Mormon Church. Key beliefs of the church were not only different from those of other Christian churches they claimed to supersede but often seemed to chastise them as well.

The details of this new religion he preached had all been revealed to him by religious prophecy and in writings engraved on golden plates that had been buried beneath hills outside of Palmyra, New York. Over the course of four years, from 1823 to 1826, he had been visited by the angel Moroni each September 21 and told of the hidden golden plates. Finally in 1827 the golden plates were delivered into his hands along with instruments known as "interpreters," the miraculous *Urim* and *Thummin*, that he would use to translate the "reformed Egyptian" on the plates into English to create the *Book of Mormon*. The text, as Smith revealed it, contained an account both sacred and secular of prophets and people who were ancestors of the American Indians. Part of a colony of Israelite origin, it was said that they had lived in the New World from about 600 B.C. to 420 A.D. In the teachings of the church, prophecies and visions were not just limited to the distant past in Israel and Egypt but took place here and now. The Garden of Eden had originally been located somewhere near Kansas City, Missouri. The head of the Mormon Church was not just a figurehead but a prophet, seer, and revelator, God's spokesman on earth. Three months before his death, as the church teetered on the edge of chaos, Smith would select a secret council of fifty who would in turn ordain and crown him as king of the kingdom of God. Several years before his murder, Smith had spoken of preparing a New Jerusalem, a place where the faithful could all be "gathered into one." Less than two years after his death, Brigham Young would lead them to Utah and the promised land.

● ● ●

In February 1846 the first wagon trains crossed the Mississippi from Nauvoo while the river was covered with ice, heading toward camps on the banks of the Missouri River where they would spend the following winter making preparations for the long trek west. By year's end some fifteen thousand people, three thousand wagons, and thirty thousand head of cattle had reached the Missouri. Cholera, fever, black canker, and malaria killed hundreds both on the march and in the camps. They saw themselves as a chosen people, however, and bore their suffering in stride. While days of good weather and good hunting were a sign of God's favor, trials and hardships were merely a sign that they were being tested.

The following spring Brigham Young headed west with a small advance party of pioneers—171 in all—to scout out the new land they had chosen to settle and look for the site of Zion. After a hard trip across the Great Plains and the Rockies they came through Emigration Canyon along the edge of the high peaks of the Wasatch Front and saw the Salt Lake Valley and the broad lake beyond it opening up before them. Others were already setting up camp and laying out fields when the wagon bearing Brigham Young, ill with mountain fever, paused at the crest overlooking the valley below. They would travel no further, he decided. This is the place. They had found the proper place, the place where they would build the promised land. Before their arrival the land here had been known only to scattered bands of Ute and Paiute and mountain men like Jim Bridger. While richer and greener land lay farther west in California and Oregon, the Mormons would elect to build their world right here on the edge of the desert. After unloading their wagons and planting their fields, some headed back across the mountains to meet the others traveling behind them and show them the way to the promised land. By year's end some sixteen hundred had arrived.

While some headed east, other scouts headed to the north, south, and west looking for new land to settle. Although the Salt Lake Valley was fertile and well watered by the standards of the Intermountain West, it had nowhere near the land to support the seventy thousand immigrants the church expected. In the deserts to the west they found only a barren landscape of salt flats and sun-baked, waterless valleys. Along the front range of the mountains that ran to the north and the deserts that ran to the south, however, they found a rich landscape watered by streams and springs that flowed from the high ground above.

In less than ten years more than a hundred Mormon settlements had spread out along the edge of the Wasatch Front and the high plateaus, reaching all the way from Bear Lake near the Idaho border to Santa Clara down toward Arizona, a distance of more than 450 miles. They came by wagon, horse, and cart. Some, like the penniless immigrants from Europe who arrived in the 1850s, made the long trip on foot. Gathered up in the docks of Liverpool, they sailed to New York and took the train west to Iowa. At the end of the Rock Island Line they loaded their possessions onto handcarts and began the long trip west to Salt Lake City on foot, pushing their hand-carts—laden with four hundred to five hundred pounds of sup-plies—in front of them, covering as much as thirty miles per day. By the time the first transcontinental railroad was completed in 1869, more than eighty thousand had made the long trek west to Zion.

The desert, or so it seemed, could absorb them all. Like the Hopi who migrated north, south, east, and west, upon their emergence from the Third World and their search for the center of things atop their high mesas in northern Arizona, the Mormons' journey to the West would become their own trail of migration and wandering. Salt Lake City became the city of Zion, the center of a new world and a new faith. Until 1850 when the territory of Utah was estab-lished, the immigrants to this desert Zion lived under the rule of a "theo-democracy," the church and state inextricably tied together.

Sharing a common faith and a common work ethic, they gath-ered together in small towns and villages. For a time church leaders dreamed of settling the desert from the Wasatch Front all the way to the edge of the Sierra Nevada in California, petitioning the federal government for admission to the Union as the State of Deseret. They established outposts in Las Vegas and San Bernardino, securing trade routes to the West Coast, and carefully guarded their isolation. While they called themselves the Latter-Day Saints, those outside their close-knit community of faith were known simply as gentiles. "We do not intend to have any trade or commerce with the gentile world," said Brigham Young. "The Kingdom of God cannot rise independent of the gentile nations until we produce, manufacture and make every article of use, convenience or necessity among our own people. I am determined to cut out every thread of this kind and live free and independent, untrammeled by any of their detestable customs and practices."

They took as their symbol not the lone cowboy but the beehive. At places like Orderville in the remote plateaus near what is now Bryce Canyon National Park, groups of Mormons experimented with communal living, setting up what were, in effect for a time, self-sufficient communes. They grew all their own food and made their own clothes and supplies, sharing equally in the colony's profits and losses. Others were less extreme, but the Mormon way of life was far different than that of any other American immigrant group. The Mormons were not rugged individualists but a determined and disciplined group. Water is in short supply in the desert, available only in limited amounts and limited places. These two elements, water and faith, would draw the Mormons together, creating a social fabric unlike that seen anywhere else in the United States.

Mormon dreams of Deseret would last no more than a few years. The close of the Mexican-American War and the discovery of gold in the Sierras would send a surge of prospectors and pioneers westward through Utah toward California. In need of food and supplies, they became a lucrative source of income for Mormon farmers and traders. Old antagonisms, however, still simmered beneath the surface, but here the Mormons held the upper hand. In 1857 a group of Mormon raiders slaughtered a party of California-bound immigrants who camped amid the high plateaus of southern Utah, an event that became known as the Mountain Meadows Massacre. The United States responded by sending troops to Utah to hold the Mormons in check. This time, however, there would be no full-fledged war, but Utah's petition for statehood would not be approved until 1896.

The Mormons would fight their own small wars with the tribes around them as well. While settlers sought to transform the dry valleys at the foot of the mountains and plateaus into gardens, missionaries tried to bring the natives to the church. For a brief time along the edge of the high plateaus Mormons and natives mixed freely. It was cheaper, Brigham Young taught, to feed the Indians than fight them, and Ute and Paiute wandered in and out of Mormon homes, joining them for dinner and sharing their food—particularly when supplies ran short in the wild. Relations, however, were often uneasy, but not without their humorous moments: Pioneer diaries record instances of natives inviting themselves in for dinner, and then watching the Mormons squirm with uneasiness as they alternately proposed marriage and jokingly threatened them with scowling faces, playing the role of the inscrutable Indian.

The sense of space that pervades the West is deceptive. It gives one the impression that resources and possibilities for life are unlimited. In reality the hunting and gathering culture of the region's native Ute and Paiute had already stretched the region's resources to the limit. As Mormon farms and villages spread, native pastures and hunting grounds shrank, and, in time, they would crowd the Indians out.

Along the edge of the high plateaus the struggle would result in the Black Hawk War. From 1865 to 1867 Mormon settlers in the valleys and bands of Ute in the high plateaus behind them lived in a state of open war. The Ute were led by a leader of the San Pitch band known as Antequer, whom the Mormons knew as Black Hawk. With food in short supply he was able to piece together a band of warriors to try and drive the Mormons out. While the Ute swept down from the high plateaus in lightning raids, the Mormon towns "forted up" and organized militias. For more than two years the two sides swapped murders and brutalities, but the natives were outnumbered and outgunned. In 1867 Black Hawk surrendered. In constant pain from old wounds, he showed up at the Uintah Indian reservation in the Uinta Valley more than 100 miles to the east and turned himself in, asking for peace. Two years later he insisted on visiting the high plateaus that had once been his people's home. Accompanied by escorts and marching bands from Mormon villages he traveled from town to town all the way from Payson to Cedar City, visiting the communities he had raided during the war. He did not just pass through but made speeches, apologizing personally for the suffering the war had caused.

The high plateaus had always been Indian land, off-limits to the towns and villages below. After the close of the Black Hawk War, the Mormons began to use the high country above them as well. They did not settle there—unlike the valleys and mountain parks in the Rockies to the east, the tops of the high plateaus were too cold and exposed for year-round farms and ranches—but used them for grazing, hunting and wood cutting—common land to be used by all.

On top of the Wasatch Plateau, Stevens said, the grass was so high that herders took to tying bells around the necks of their cattle and oxen to keep from losing them in the tall grass. Sawmills appeared in the groves of pine and fir and began turning out lumber for the growing farms and towns below. The floor of Ephraim Canyon, a boulder-choked stream today, was a sandy wash, so smooth that one could drive a horse and buggy along it all the way to the top of the plateau. Hunters ventured up the canyons and into the forests above and came back down with wagonloads of deer.

Sheep began appearing as well as herders wintered their flocks in the western deserts and drove them up on top of the plateau in summer. With the railroad running through nearby Soldier Canyon, transient herds from outside the area began appearing too—some from as far away as Oregon. By the late 1800s there were more than eight hundred thousand sheep grazing on top of the plateau. If a young man were diligent, old-timers used to joke, he could go up the plateau with two hundred head of sheep in the summer and come back down with more than a thousand by fall—collecting the "strays" of others.

Like the desert grasslands below, however, these high-altitude meadows had not evolved under heavy grazing pressure—and the results were disastrous. In less than two decades it was all gone. High stands of grass on top of the plateau were replaced by fields of broken rock and loose dirt. With grass in short supply, herders began following the retreating banks of snow up the flanks of the plateau in the spring, cropping the grass and the plants before they ever had a chance to grow. Standing in town, it was said, you could tell how many bands of sheep were moving across the top of the plateau by counting the dust clouds moving through the sky.

Overgrazing had its impact on the towns below as well. Stripped of its cover of grasses, soils and rocks from higher up began to wash down below. Logging left slopes and hillsides bare, triggering landslides that sent boulders and stones tumbling into the canyons below. With no plant cover the land lost its ability to soak up water as well. Streams that once ran year-round suddenly ran dry during the long, hot days of summer. In the spring, streams like the one that ran through Ephraim Canyon spawned mudslides and debris flows—carpeting city streets and fields with mud and rock. The water reaching the town was often fouled as well by herds of cattle and sheep and unsafe to drink. "Experts at the time claimed that streams purified themselves after flowing five or ten miles," Stevens said. "But once you start finding sheep shit in your water you know there's a problem no matter what the experts say." Once a fruitful garden, the land at the base of the high plateaus was being transformed into a cesspool.

Mudslides and flooding became a problem not just at Ephraim but up and down the length of the high plateaus and throughout much of the rest of the Intermountain West. Overgrazing and clear-cutting had stripped the mountains bare, damaging not only the peaks but land below them as well. The early Mormon settlers were not like most other pioneers, Stevens said, in the fact that most, before coming here, were not farmers or ranchers but city people with a strong interest in both science and academics. When problems arose they began looking for solutions, he said.

The nearby town of Manti was the first to take action on the growing water pollution and flooding problem. With an eye toward controlling grazing, in 1901 they incorporated the entire watershed that supplied the town right up to the crest of the plateau and closed it to grazing—and hired a watchman to shoot trespassing cattle and sheep on sight. With the livestock gone the water cleared. Flooding stopped almost immediately. A year later the town's mayor traveled to Washington, D.C., and pleaded with President Teddy Roosevelt and Gifford Pinchot, the father of the U.S. Forest Service, to give the town forest protection as a national forest. A year later the Manti Forest Preserve became Manti National Forest.

Elsewhere, including Ephraim, the floods continued. Ranchers and herders tried to shift the blame to others or changes in nature, but by 1909 written reports were identifying overgrazing as the culprit behind the region's flood problems. Searching for solutions, the U.S. Forest Service set up the Utah Range Experiment Station in

Ephraim Canyon to study the region's flooding problems. Climbing up the flanks of the plateau above Ephraim Canyon, the study area reached all the way from an elevation of 6,800 feet amid the pinyons and junipers above the base of the plateau to an elevation of 10,300 feet near its crest—spanning a broad range of forests and climates.

Working on a shoestring budget, they did the first comprehensive watershed study in the United States and proved the link between overgrazing and flooding—documenting beyond any shadow of a doubt that barren, overgrazed land failed to hold either soil or water in place. Soon they began looking for ways to restore the abused land as well: experimenting with seedings, plantings, terraces. From there their work spread out to include restoring bits and pieces of the habitat for wildlife and studying the response of plants to grazing to get a better idea of the carrying capacity of the land—the number of cattle and sheep that could be safely grazed in the area while keeping the land healthy and intact. Timing was critical as well: learning when the plants set seeds, when they were fragile, when they could be safely cropped.

Ultimately, Stevens said, they took grazing and land management in the West out of the realm of guesswork and pushed it into the realm of science. A generation of land managers throughout the West were trained at the Experiment Station in Ephraim Canyon—although the site is largely forgotten today. The station, in turn, worked closely with local farmers and ranchers, getting their comments and input and taking them on field trips to show them the projects they were working on and explain their findings.

The work at the Experiment Station both brought flooding under control and improved grazing for both livestock and wildlife. Today flocks of sheep graze on the tops of the high plateaus. Herds of deer and elk and other wildlife thrive here. Streams flowing down their flanks still provide water for towns and fields beneath them. Restoring the land would not come easy, but those who lived here would learn a hard lesson from all of this: The land here is as fragile as it is rugged. The damages that followed the rapid settlement of the West are ones we are still trying to repair.

More than a century after the first pioneer's arrival, the Mormon Church still dominates the religious, political, and social life of Utah. Mormon missionaries still travel around the world in search of converts just as they did in the late 1800s. For young men active in the church, "going on mission" is still a rite of passage. Changing planes in Salt Lake City on the way to Chicago, Seattle, or Los Angeles, you can see crowds of family and friends waiting to greet the returning missionaries on their way back from a two-year stint in such faraway places as Russia or India. They gather around arrival gates and security checkpoints bearing signs such as: "Welcome Home Elder Burton!"

Back home, however, the Mormons themselves are no longer a collection of utopian farmers living in isolated, self-sufficient towns and villages, but an urban and suburban people whose economic life is little different from those of the non-Mormon world around them. They work in offices and factories and buy their food and clothes at supermarkets and shopping malls. More than half of the state's population is concentrated in the urban sprawl of the Salt Lake City area, tucked up tight against the high peaks of the Wasatch Front, not far from the edge of the Great Salt Lake. Tourism is one of the state's leading industries. While national parks out on the Colorado Plateau draw tourists by the millions over the long, hot months of summer, the ski resorts scattered along the mountain front just outside of town bring in several million over the winter.

Others, of course, still farm the land, but less than four percent of the state's roughly 85,000 square miles is suitable for farming—and almost all of that is already under cultivation. Farming and ranching is still as precarious here as it was when the first Mormon pioneers arrived more than a century ago. But while the first Mormon pioneers labored long and hard to bring their development in line with the limits of the landscape here, their descendants often seem determined to push it as far as possible—intent on unraveling and dismantling the revolutionary ideas of land management, the carefully planned control of grazing, logging, mining, and water pollution, that their ancestors worked so hard to develop. The early church often fostered that early conservation ethic as well. The land here was sacred, the site of the New Jerusalem. Instead of champions of conservation and wise use, today far too many of the state's farmers, ranchers, and politicians have become champions of overdevelopment and abuse, more concerned about the short-term bottom line than the long-term health of the land and water they depend upon for their survival. Their attitudes, of course, are no different from those of many elsewhere in the United States: Ohio, New York, California, or Florida. They are surprising only in light of their past—the strong sense of place that characterized the lives of early pioneers here, their understanding of the rhythm and pattern of things around them.

Pioneers, of course, clear the ground not only for themselves, but for others who will quickly follow them. The Mormon's long-sought isolation lasted only a few decades. Land rushes would soon bring others through Utah and eventually to it. Like the Native and the Hispanic peoples before them, they too would soon find themselves surrounded by strangers. After the Second World War, the rest of the United States would wash across the Colorado Plateau like a wave.

BOOK SIX

Center

IN THE EVENING, about supper-time, feeling somewhat guilty and contrite—for they are, most of them, really good people and not actually as simple-minded as they pretend to encourage me to pretend us all to be—I visit them again around the fires and picnic tables, help them eat their pickles and drink their beer, and make perhaps a trace of contact by revealing that I, too, like most of them, come from that lost village back in the hills, am also exiled, a displaced person, an internal immigrant in this new America of concrete and iron which none of us can quite understand or accept or wholly love. I may also, if I am lucky, find one or two or three with whom I can share a little more—those rumors from the underground where whatever hope we still have must be found.

EDWARD ABBEY
Desert Solitaire

TWENTY·ONE

The Well-Traveled Wilderness

IT IS EARLY September, and high on the flanks of the Abajo Mountains the aspen groves are already beginning to turn: shades of bright gold against the surrounding dark forests of spruce and fir. Up above the trees are meadows of knee-high grass, still green from late-summer rains. At sunrise the ground was covered by frost. By late afternoon the day had turned warm and bright.

I had spent the previous night camped several thousand feet below on the edge of the desert, on top of a high mesa at the base of the mountains. I had fallen asleep while the moon was still above the edge of the horizon, listening to the barks and howls of coyotes prowling the canyons below. The following morning I worked my way up into the peaks by degrees, following dirt roads and trails. On the edge of a high meadow I surprised a bull elk grazing in the tall grass. Sprinting down the mountain, he disappeared so quickly into the trees that I began to wonder if I had even seen him at all. No trails led upward above the trees. I climbed the last thousand feet cross-country, up through the meadows and across the steep slopes of boulder fields, pausing from time to time to catch my breath and take in the rapidly expanding view.

On top was a windblown forest of stunted evergreen trees, the tallest no more than head high. Walking across the rounded summit of the peak you could see for miles between the thin, scattered trunks of the trees, across the surrounding deserts to the lines of distant plateaus and peaks on the edge of the horizon.

The Abajos lie almost squarely in the center of the Colorado Plateau. From their crest you can look north beyond the even higher rise of the La Sal Mountains and see the curving wall of the Book Cliffs that run through central Utah. To the south you can see the blue waters of Lake Powell and the solitary spires of rock that rise up from the floor of Monument Valley. To the east and west the views reach even further: past the tilted rise of Mesa Verde a hundred miles to the east all the way to the snow-covered peaks of the San Juan Mountains in Colorado and beyond the solitary rise of the Henry Mountains to the west all the way to the high plateaus of the Wasatch Front.

In between these distant points of reference are slickrock deserts of brightly colored stones: a tangled maze of canyons, mesas, and buttes that stretches for miles in every direction, the red-rock heart of the Colorado Plateau. Scattered across it are green forests of pinyon and juniper and khaki-colored deserts of dry grass and sage, but what catches the eye are the brightly colored rocks that climb up the faces of the canyons and cliffs below: improbable shades of red, white, and brown. In places there are even purple and pink layers of sandstone and shale. They run through the landscape in flat, even layers, like the shades of color in a serigraph.

Beyond the base of the mountains to the north lies Canyonlands National Park, a labyrinth of canyons cut into the slickrock by the converging waters of the Colorado and the Green. The rivers' junction is hidden from view by successive walls of rock. Here and there in the heart of the canyons fins, spires, and arches of rock rise up out of the ground like the ruins of an ancient city or fortress. The Green River begins several hundred miles to the north in the Wind River Mountains of Wyoming; the Colorado starts several hundred miles to the east, not far from the front range of the Rockies that overlooks the Great Plains. Mountain streams, they rise in power and strength as they head downhill toward the sea, transformed into rivers by the time they reach the desert, cutting through the finely layered rocks of the Colorado Plateau in deep canyons.

After joining with the Green, the Colorado River cuts even deeper into the rocks, heading south toward Lake Powell between the three-thousand-foot-high walls of Cataract Canyon that lies just west of the

Abajos. Streams flowing down the western flanks of the mountains feed into rugged, winding side canyons: Dark, Gypsum, and No-Name, which run all the way to the river. Farther south the Colorado is joined by other rivers as well—the San Juan, San Rafael, Escalante, and Paria—growing in size and strength as it heads toward the mile-deep reach of the Grand Canyon in northwestern Arizona. It is not just the rivers that have carved the landscape here but the streams and washes that feed into them as well. Flying over the Colorado Plateau at thirty thousand feet on a flight from Los Angeles to Chicago or New York, you can trace their paths for miles, twisting and winding through the barren landscape below like the tendrils of a vine.

While the rocks they slice through are often several hundred million years old, the canyons themselves are often no more than 10 million years old and quite possibly less. Sometime between 20 and 5 million years ago, when the Colorado Plateau began its rise, the rivers and streams that ran across it began to cut down into the ground, etching the once flat surface of the land here with literally thousands of canyons and cliffs. The process is still continuing today. Following the deep cuts of rivers and streams, the cliffs here are steadily retreating, widening canyons into broad desert valleys and turning mesas into buttes, peeling back the overlying layers of rock to reveal the details of the ancient landscape buried beneath them.

In other parts of the world desert rivers are sources of civilization and life: the Tigris and Euphrates of the Middle East; the Nile of Africa. Here, however, they are barriers to travel and communication. Flowing in inaccessible canyons, they divide the vast reach of the desert here into a collection of isolated mesas and plateaus. Although they carry the collected runoff and rainfall from several thousand square miles of mountain and desert, they are not bordered by broad, fertile floodplains but high walls of rock. The rivers themselves are so choked with rapids that they are all but impassable, navigable only in specially built boats and rafts. They emerge from their canyons only sporadically in places like the Uinta Valley where the Green River meanders through flat open ground, or near the Uncompahgre Plateau in western Colorado where the Colorado winds through the level reach of the Grand Valley that lies beyond the Rockies. The banks of the rivers here are surrounded by thin ribbons of green: galleries of cottonwood trees, narrow meadows of tall grass, and irrigated fields of corn and alfalfa.

In spite of its ruggedness and isolation, the land here has been inhabited for centuries. In the canyons that spread out from the flanks of the Abajos you will find the ruins of small pueblos and panels of pictographs from prehistoric peoples like the Fremont and Anasazi. Ute and Navajo passed through here as well. Today their reservations lie only a few miles away. Streams flowing down from the high peaks here carry water to Blanding and Monticello, towns founded by Mormon pioneers in the late 1800s. The first to arrive did not volunteer to settle the rugged and inaccessible landscape below but were ordered into service by their church, commanded to settle the canyons and deserts of the Colorado Plateau to help expand the borders of Zion. Most had fled the East and Europe only a few years before. Having reached the promised land and built up a new life in the shadow of the high plateaus farther west, they gave up homes, farms, businesses, and ranches and loaded their belongings into wagons and carts and headed off into the canyons and slickrock deserts that lay behind their backs. Some were ordered southward into the rugged and unknown world that lay to the north of the Grand Canyon. Others were sent east to settle along the San Juan and the upper reaches of the Colorado. The first wagon trains headed straight across: up and over the San Rafael Swell, through the canyons of the Escalante, Green, and Colorado, finding trails and building roads as they went. In the smooth rock walls of Glen Canyon they carved out a route with pickaxes, hammers, and dynamite, then locked their wheels with chains to brake their descent and lowered them over the side. Floating their wagons and herds across the river, they then dug their way back up the other side. They called the crossing Hole-in-the-Rock. Years later, survivors still wept when they recalled the trip. Later arrivals soon learned to take the longer and flatter route through the San Rafael Desert to the north, following the Colorado and the Green southward into the slickrock.

While settlers managed to carve a life out of the rock, the settlements here never thrived like those to the west on the well-watered flanks of the high plateaus and peaks of the Wasatch Front. For those who followed the Mormons to the Intermountain West, the Colorado Plateau and its deep canyons were an obstacle to be avoided. Booms in fur, gold, and land sent thousands of pioneers streaming across the West but left the Colorado Plateau behind. Only the cattle boom of the late 1800s seemed to take hold here, but then only briefly. Overused, the virgin desert grasslands quickly disappeared. Instead the Colorado Plateau became a place of refuge and escape, its canyons a hideout for bands of horse thieves, cattle rustlers, bank robbers, and murderers like Matt Warner, Elzy Lay, Billy McCarty, and a rebellious Mormon farm boy by the name of Charles LeRoy Parker— who found that it was far easier to steal a fortune here than make one. Later known as Butch Cassidy, he would team up with Harvey Longabaugh—the Sundance Kid—to head a loosely organized band of predatory thieves known as the Wild Bunch. They settled not far from here in the midst of a maze of canyons bordered by the Colorado, Green, and Dirty Devil rivers a few miles west of the Abajos, an area that would eventually become known as Robbers Roost. Their robberies and raids reached as far north as Montana and as far west as Nevada. Stealing horses, cattle, and mine payrolls from places like Telluride and Castlegate, they would speed across the desert and lose their pursuers in the canyons. Few knew the land well enough to find them or were well-paid enough to risk tracking them down.

When John Wesley Powell made his historic trip down the Grand Canyon in 1869, the Colorado Plateau was *terra incognita*, a blank space on the maps marked not by symbols and signs but by a single word: unexplored. Today it is a landscape in harness. Its rivers and streams have been dammed to provide cheap water and power to sprawling desert cities like Los Angeles, Phoenix, and Las Vegas and to irrigate improbable fields of corn and rice in the even hotter deserts to the south—not agriculture but agribusiness, corporate farms that sprawl over thousands of acres. Beyond the reach of the dams and their artificial lakes, strip mines feed coal from the deposits of ancient seas to giant power plants at places like Page, Arizona, and Farmington, New Mexico. In the early 1970s astronauts en route to a landing on the moon reported that they could see smoke from the Four Corners Power Plant near Farmington as they left the earth's orbit—one of the few signs of human life visible from space.

Although the land here is rugged, it is also fragile. For the rivers, the proliferation of dams has changed almost everything about them: not only their rates of flow and sediment loads but their temperature and ecology as well, pushing native species of fish like the hump-backed chub and the Colorado squawfish to the verge of extinction. Water here no longer flows downhill but in the direction of power and money—or so the skeptics of the region's so-called development say.

World War II changed the region forever. In 1945 scientists working at Los Alamos, New Mexico, developed the first atomic bomb. Unleashing the hidden power of the atom created a sudden need for uranium and other radioactive elements. As the Second World War gave way to the cold war, miners and prospectors flocked to the Colorado Plateau, prospecting not with picks and pans but Geiger counters, looking for seams of hot, radioactive rock. While finds of silver and gold proved elusive here during the earlier mining booms in the West, the ground here was shot through with seams of uranium ore and other radioactive rocks. While Mormon pioneers looked for rivers and streams to supply them with water, uranium miners looked for the paths of ancient rivers and streams in the brightly colored sand-stones and shales of the Chinle and Morrison formations. Eruptions of volcanic ash and lavas around the perimeter of the plateau and pos-sibly the intrusion of the region's laccoliths had laced the groundwater here with trace amounts of uranium and other radioactive elements. Out in the desert, rainwater percolating down through the rocks above gathered into the channels of ancient rivers and streams buried within the rocks below and began flowing beneath the desert's sur-face. Where the groundwater running through these buried channels encountered oxygen-starved shales—typically the deposits of ancient marshes and lagoons laden with the debris of plants and trees—the uranium was deposited and filtered out like gold in a swirling pan of silt-laden water.

While early Mormon pioneers had marked out scattered tracks and trails or used those of the Spaniards and native peoples who had preceded them, uranium miners laced the deserts here with gravel roads and jeep trails that led out to mines and claims on top of remote mesas and halfway down the walls of rugged canyons.

Almost nothing was left untouched or unexplored. Cover a detailed map of the region today with a square-mile grid and you will be hard-pressed to find a single square mile not reached by a bone-jarring four-wheel-drive road or trail of some kind—most of them built by uranium miners in the 1950s. The close of the war not only created a sudden demand for uranium but flooded the markets with equipment: surplus army bulldozers, trucks, and jeeps, vehicles that made the impassable terrain here suddenly accessible—or at least far more accessible than it had ever been in the past.

The uranium boom lasted little more than a decade or two. By the late 1970s, however, tourism in the region was beginning to grow. Places like the Grand Canyon and Mesa Verde had been popular since the turn of the century. Now, however, backpackers and rafters flock to the region in search of wilderness, and began traveling into the backcountry in increasingly greater numbers, following the roads and trails of uranium miners into the canyons and desert.

Today it is a landscape under siege. In the past decade tourism in the region has exploded. At the height of the summer tourist season, lines of cars at the entrance station to the South Rim of Grand Canyon National Park reach back for more than a mile. Fistfights erupt over parking spaces at lodges and stores. Walk out to the edge of one of the overlooks on the canyon rim and you will find yourself standing shoulder to shoulder with hundreds of others. Trips into the backcountry bring relief from the crowds, but require months of advance planning to secure a reservation. The waiting list for private rafting trips through the canyon is now nine years long. Little more than a century ago it was an unknown and uncharted wilderness. The story is much the same over the long, hot months of summer at Zion, Bryce, and Mesa Verde. Over in southeastern Utah, Lake Powell alone draws nearly 3 million visitors per year. Farther north, even Canyonlands National Park, possibly the most remote and undeveloped National Park in the United States outside of Alaska, is beginning to show signs of strain. Backcountry use in once-remote reaches of the park like the Maze district and the Needles district has become so heavy that the park service has been forced to install pit toilets alongside trails and is in the process of putting together a computerized reservation system for travel, parceling the wilderness out like places on a package tour. What was once wild and remote has become a name-brand stock item, featured in coffee table books and wilderness engagement calendars.

While Anasazi farmers, Spanish colonists, and Mormon pioneers sought out fertile, well-watered pockets of ground to plant their fields and build their villages and towns, we seek out the extreme and exotic: the deepest canyon, the highest arch, the hottest desert. In the 1960s and 1970s the writer Edward Abbey published a series of eloquent essays mingled with angry polemics in books like *Desert Solitaire, The Journey Home, and The Monkey Wrench Gang,* extolling the unknown beauty of the deserts here and railing against the development that seemed poised to destroy forever—fantastic plans for highways, strip mines and dams criss-crossing the still-wild reach of the Colorado Plateau. Today, however, the biggest threat to the region is not industry or ranching but tourism—a steadily increasing flood of visitors who are straining the parks and the wilderness areas around them to the breaking point.

The signs of damage and overuse are everywhere. They range from archaeological sites stripped clean by pothunters and visitors in search of souvenirs to trampled cryptobiotic soils flattened by mountain bikers and hikers out exploring the desert. We have, in the words of the cartoon strip character Pogo, met the enemy and he is us.

While rivers and streams have carved deep canyons here, the increasing flood of outsiders has steadily chipped away at the region's traditional cultures and people—Native, Hispanic, and Mormon. The world here is isolated no more. While tourists come and go, others come to stay—well-heeled refugees and retirees from Los Angeles, Denver, New York, and Dallas—buying up homes and ranches around once small towns like Durango, Santa Fe, Sedona, and Moab, permanent tourists in search of an outdoor "lifestyle" of mountain biking, skiing, rafting, and hiking. Although these new-comers have provided the region with some much needed sources of income and diversity, they have often left locals feeling increasingly inundated and isolated—priced out of their communities and crowded off land they had once thought of as their own. Beneath this thin veneer of prosperity are low-paying jobs in motels and restaurants with salaries that fail to keep up with the soaring prices of rent and real estate. Instead of getting ahead, locals often find themselves falling farther and farther behind.

In spite of the steadily increasing crowds, however, the total population here is still small. Few towns have populations of more than ten thousand and they are often separated by dozens or even hundreds of miles. In spite of the vastness of the landscape here, that sense of space is at once its most defining and most fragile aspect. While you can quickly lose sight of a hundred people in a hundred acres of hardwood forest in the east, here you can see things moving across the ground from several miles away: a hiker on the rim of a canyon, a pickup truck speeding down a dirt road. They stand out and demand attention. The landscape here is defined by a series of deficiencies: without water, without roads, without people. Put people into the landscape and you suddenly change its character entirely. Adding to the fragility of space is the nature of the landscape itself. While the land here is open, much of it is also impassible. Roads and trails here gather people the way washes and canyons in the desert gather rainfall. Problems become focused and magnified beyond all reasonable sense of proportion and scale.

In spite of our increasing numbers and abuses, the land here is still wide and open, but it is disappearing fast. Wilderness has played an important part not only in the history of its native peoples, but more recent immigrants as well. How much of that wild space we will be willing to save for those who come after us is a troubling and complex problem. Like those who lived here before us, we are still struggling to understand the limits of the landscape around us. Finding the balance between man and nature here is an elusive and ongoing struggle.

T W E N T Y · T W O

Canyons

IN 1869, two weeks after the first transcontinental train rolled through Green River, Wyoming, on the tracks of the Union Pacific, John Wesley Powell began his historic float trip through the canyon country of the Colorado Plateau. A group of ten, they set off down the Green River in four specially made oaken boats: the *No-Name, Kitty Clydes Sister, Emma Dean,* and the *Maid of the Canyon*. On board were food and supplies for a journey of ten months. From Wyoming they planned to follow the Green River south to its junction with the Grand, as the upper reaches of the Colorado River were known in those days, and then follow the Colorado River southward through the Grand Canyon, a journey of nearly a thousand miles. Ahead of them lay the last unexplored wilderness in the continental United States. South of the tracks of the Union Pacific, every map was blank.

No one, least of all Powell himself, knew exactly what they would find amidst the canyons, particularly farther south when they reached the Colorado. Some said the river contained falls higher than the Niagara or that even flowed underground. Others claimed it was a placid stream, flanked by fields of wild wheat. Whatever lay ahead, Powell planned to carefully study it. As they drifted through the canyons, they plotted their location by the stars—taking readings at night with a sextant like mariners traveling at sea. They measured changes in altitude with fragile barometers and plotted the steady drop of the river. From time to time they measured the height of the cliffs above them as well, climbing up out of the canyon, passing their barometers from hand to hand as they inched their way up through the rocks, following chimneys and cracks.

The decades that followed the Civil War were an era of exploration and discovery in the American West, a time for measuring and taking stock of the potential and possibilities of the land that lay beyond the Mississippi. From Canada to Mexico a series of government-sponsored geologic and geographic surveys ranged across the western United States, led, for the most part, by prominent scientists and military men: Clarence King, Ferdinand V. Hayden, and Lt. George M. Wheeler. Powell's own qualifications had far more to do with enthusiasm than experience. An ex-major of the Union army, his formal education had ended with high school. Beyond that he was largely self-taught, traveling widely on fossil and plant collecting trips in Wisconsin and along the Mississippi River. After the war he managed to secure an appointment as a professor of geology at Illinois State Normal University. As his abilities and qualifications grew, his interests shifted steadily westward, taking him on trips through the Great Plains and the Rockies, before finally focusing on the Colorado Plateau.

For his trip down the river he sought out support wherever he could find it, funding from the Illinois Natural History Society and supplies from the United States Army. Most of the expedition's expenses, however, were paid out of his own pocket, and they were not insignificant for a man of limited means. His group itself had been pieced together from a collection of volunteers—friends, relatives, and enthusiastic students, as well as a few mountain men hired for their ability to hunt. Included among them was Powell's brother, a shell-shocked victim of the Civil War. Powell himself had lost his right arm in the Battle of Shiloh, an obstacle some no doubt would have considered insurmountable for one intent on traveling through a wilderness of canyons and cliffs.

Powell designed the boats himself and had them built in Chicago and shipped out west by rail. The expedition and their gear created quite a stir in the booming railroad town of Green River. When

Powell and his men finally pushed off from shore after several weeks of packing and preparation, the town's residents, a few hundred in all, gathered along the shore to see them off. Publicly they wished them well. Privately they placed bets on their chances of making it out alive.

Rounding a bend on the river they drifted out of sight, into the canyons of Wyoming and northern Utah that mark the upper reaches of the Green. For the next thirty-eight days they saw no one outside of their own small group. Following the river southward, they passed through Red Canyon, now submerged beneath Flaming Gorge Reservoir, and into the Canyon of the Lodore. At Disaster Falls they lost one of their boats—the *No-Name*—and a good portion of their supplies. Salvaging what they could from the wreckage, they took to the river again, floating through Echo Park and Island Park near the mouth of the Yampa River before finally passing through the center of Split Mountain and into the broad open reach of the Uinta Valley. Once out of the canyons, they would lose one of the group as well, a volunteer who had seen all he needed of canyons and rivers on the rugged start of the trip and elected to head back east.

At the newly established Uinta Indian Agency they replaced their losses as best they could and learned that the newspapers had been filled with reports of their demise. One of the most vivid accounts came from a gifted liar and storyteller by the name of John A. Ridson, who parlayed his tale of the group's disastrous end into a free railroad ticket to Illinois. Not content to merely report their death, he made himself part of the expedition as well and increased its size from ten to twenty-five, complete with a list of fictional characters, including a half-breed guide by the name of Chick-a-Wanee. Geography of the region was fast and loose in those days, and Ridson proved as adept at manipulating the landscape as he did personnel, creating imaginary rivers like the Big Black and the Deleban as well as nonexistent towns—a large Indian settlement along the banks of the Green River with the improbable name of Williamsburg. Powell and the rest of the group, Ridson said, had perished while crossing the river in an "Indian yawl," sucked into a whirlpool while he watched helplessly from the shore. As the boat sank from view, Powell stood stoically in the stern and said, "Goodbye John! You will never see us again!"

The press bought Ridson's story wholesale. (It was so dramatic that Ridson often broke into tears relating it.) Eventually he met with the governor of Illinois to tell him personally of the disaster. In time he was denounced as an impostor by Powell's wife, Emma. None of those mentioned in Ridson's imaginary crew of twenty-five had been members of her husband's group, she explained. Then, too, there was the not-so-subtle question of timing. While Ridson claimed that Powell and his men had died on May 8, the last letter she had received from her husband was dated May 22—mailed two days before they had left Green River, Wyoming, and started down the river.

After a few days of rest and letter writing, they headed down the river again. Rougher water lay ahead. Up ahead the land seemed to rise up around the river. It was something they would notice over and over again as they headed south. Rather than flow around mountains and plateaus, the rivers here often seemed to flow right through them. Troubles returned almost as soon as they returned into the canyons. Finding their way through the rapids-choked reach of Desolation Canyon, they ran into headwinds so strong that they were all but stopped in their tracks, their boats unable to make headway even when moving with the current. In places the rapids were so rough that they were forced to portage or line their heavy boats around them—carrying them alongshore or leading them through the white water with ropes; backbreaking and often dangerous work that left them exhausted. Elsewhere they fought their way through. Breaking their oars in the rapids, they fashioned new ones out of driftwood. Past the Book Cliffs they floated out of the canyons and into the San Rafael Desert. The desert brought much needed relief from the fast-moving water of the canyons behind them, but the heat was relentless. After a few days of smooth, calm water, they headed back into the rock, following the meandering path of the Green River southward between endless ranks of mesas and high plateaus.

They reached the mouth of the Grand on July 17. The trip was now more than halfway through. The Colorado and the Grand Canyon still lay ahead. They camped out near the rivers' junction and took stock of their provisions. Losses and spoilage had reduced their ten-month supply of food to no more than two. Their plans for a slow and detailed study of the even deeper canyons ahead were beginning to unravel. The danger of their situation, however, did not entirely overshadow their appreciation of the landscape around them. Their camp lay in the midst of what is now Canyonlands National Park. Trying to get a sense of where they were, Powell and a partner from among the group climbed up to the canyon rim. Years later Powell would lovingly describe the view:

> Below is the canyon through which the Colorado runs. We can trace its course for miles, and at points catch glimpses of the river. From the northwest comes the Green in a narrow winding gorge. From the northeast comes the Grand, through a canyon that seems bottomless from where we stand. Away to the west are lines of cliffs and ledges of rock—not such ledges as the reader may have seen where the quarryman splits his blocks, but ledges from which the gods might quarry mountains that, rolled out on the plain below, would stand a lofty range; and not such cliffs as the reader may have seen where the swallow builds its nest, but cliffs where the soaring eagle is lost to view ere he reaches the summit. Between us and the distant cliffs are the strangely carved pinnacled rocks of the Toom'pin wunear' Tuweap. On the summit of the opposite wall of the canyon are rock forms that we do not understand. Away to the east a group of eruptive mountains are seen—the Sierra La Sal which we first saw two days ago through the canyon of the Grand. Their slopes are covered with pines, and deep gulches are flanked with great crags, and snow fields are seen near their summits. So the mountains are in uniform,— green, grey and silver. Wherever we look there is but a wilderness of rocks,—deep gorges where the rivers are lost below the cliffs and towers and pinnacles, and ten thousand strangely carved forms in every direction, and beyond them mountains blending with the clouds.

By the time they started down Cataract Canyon, both their flour and bacon had started to spoil. Most of their clothes and equipment had been lost to the river, and what remained was wearing thin. They spent the nights on sandbars and beaches along the river as they had done before, but the novelty had long since gone. They found their camps filthy with dust and alive with insects. On more than one occasion their cooking fires set thickets of brush and trees alongside them on fire and sent them rushing headlong into the river to escape the flames. There were no guides, logs, or even maps to guide them. No one—Indian, Spanish, or Mormon—had been down the river before them.

As they floated down the river they named passing canyons, cliffs, and streams: the Dirty Devil River and Bright Angel Creek; Music Temple; Glen and Marble canyons. Each reach of the river was different. In Glen Canyon the river was still and quiet, weaving its way through soft, smooth layers of red, white, and pink sandstone. Hidden in alcoves and grottoes carved into the rock were hanging gardens of ferns and flowers and hidden forests of pine and spruce, even redbuds and dogwoods. Beyond it in Marble Canyon the river was swift, flowing between layers of hard, dark rock, polished until they gleamed like glass by silt-laden sheets of water that tumbled down from the desert above when the dry land was drenched with rain.

They spent three weeks traveling through Cataract, Glen, and Marble canyons, reaching the mouth of the Little Colorado on August 10. Up ahead was the start of the Grand Canyon. Two and a half months of river travel had taught them that the character of the river changed dramatically in response to the rocks around it: gentle and slow where it was fringed by soft layers of sandstone; hard and fast where it flowed through hard layers of limestone. Four days later as they headed into the Grand Canyon they watched in fear as the hard, black rocks of the Inner Gorge rose up around them and the river narrowed. Up ahead was a nightmare of water and sound. The rapids never seemed to end, each one worse than the last. Pinnacles and crags of hard, black rock rose up out of the water to either side: twisted layers of Vishnu Schist shot through with jagged veins of bright, white rock. Unable to portage or line their boats through, they shot the rapids as best they could.

By August 25 they had reached Lava Falls, the worst the canyon had to offer. They were more than halfway through the canyon—although they did not know it at the time. Their boats were a wreck, nearly falling apart. Three months before they had drifted through still stretches of water with their boats lashed together, Powell reading poetry aloud to his men: snatches of Tennyson and Sir Walter Scott. Now they were on the verge of mutiny. What had started out as a carefully planned expedition had become a race for survival. Three days later they reached Separation Rapids. Their remaining flour and bacon had become so rancid that they were forced to throw it away. What was left would last no more than five days. Dazed by the trip and awed by the rapids ahead, they decided to camp overnight and think it over. In the morning three of Powell's men decided to walk out on foot rather than risk the rapids. There were no threats or recriminations. They had all been pushed to their limit. Lost a mile deep in the earth in the midst of a vast desert, neither group was sure that they would ever make it out alive.

They divided their food and supplies and left one boat behind in case those heading out on foot changed their mind. In a matter of minutes it was all over. Powell and his remaining men ran the rapids safely—for once the rapids had been far easier than they looked. Reaching the still water beyond, they fired off their pistols in hopes that the others would follow, but there was no reply. Four days later they passed through the Grand Wash Cliffs that mark the western edge of the Grand Canyon and drifted into the broad flat deserts beyond. The next day they passed the mouth of the Virgin River and found four men fishing along the river, a Mormon by the name of Asa and his two sons along with an Indian companion, casting their nets out over the water. Their meeting was no accident. Brigham Young had sent word out through his church, telling those of his flock who had settled in the south to watch the river for signs of wreckage or bodies, the remains of the Powell Expedition, once again rumored to have vanished in the unknown canyons upstream.

Four of the remaining six would continue down the river. Two would travel all the way to Mexico and the river's mouth in the Gulf of California, where the Colorado finally reaches the sea. Powell and his brother, however, left the river for the Mormon settlement of Saint Thomas and then pushed north toward Salt Lake City, seeking news of their companions who had left the river at Separation Rapids and were attempting to cross the desert on foot. Mormon scouts carried the message from town to town and scoured the desert for tracks and signs, but Powell and his brother were already heading back East by train by the time their missing companions had been run to ground. They had been found dead on a stretch of bare rock in the desert near a *tinaja*, or pothole, a temporary pool of water, stripped naked, their bodies shot full of Shivwits arrows. It was simply bad luck. They had been in the wrong place at the wrong time. The Shivwit warriors who killed them had been looking for a group of trappers who had shot and raped an Indian woman. The following year Powell would actually track down the band that had murdered his friends and sleep alongside them unarmed. Meeting Powell they apologized. "We are sorry," they told him. "If we had known they were your friends we would not have killed them." Powell's three men, they realized, were not the ones they had been looking for. But the year before when the Indians had pressed Powell's lost companions to explain themselves, they had told them simply that they had come down the great river below by boat and their story was not believed. No one who ventured down the river, they knew, came out of the canyon alive.

Powell's trip down the river transformed him into a celebrity. Traveling around the country on the lecture circuit, he brought the canyon country to life in the national imagination—thrilling his audiences with stories of adventure and descriptions of a landscape unlike any most of them had ever seen. The following year he was back on the river—better known and better funded, but his center of interest had shifted to other things. Over the course of that second trip he was continually on and off the river, shuttling back and forth between the Southwest and Washington, D.C., as he looked for funding and political support for future projects, developing plans for a systematic scientific survey of the entire region.

Running the river was merely the beginning. The Colorado Plateau was not only a geographic frontier but a scientific one as well: full of facts, features, and peoples unlike those seen anywhere else on earth. Powell's river trip would become the seed for a systematic survey of the entire region, one that continued for several years—mapping the course of rivers and canyons, defining the dimensions and limits of mountains and plateaus, giving a definite size and shape to the landscape they had revealed to the outside world.

Living and working in the region, they would develop close ties to both Mormons and Indians—contacts that would shape much of Powell's later work. Native cultures in the West at the time were still largely untouched, but they were also rapidly disappearing. Contact with Europeans and other American immigrants, Powell knew, would change them forever, and he soon became as obsessed with documenting the details of their cultures, customs, and languages as he had been about running the river. While other surveyors and explorers working in the region seldom traveled far without military escorts and protection, Powell and his men often traveled alone and unarmed. While he worked at unraveling the details of the region's geology and geography, he learned native languages as well, becoming fluent in Ute, Hopi, Paiute, and Shoshone. In the decade that followed his trip down the Colorado River, Powell would play an instrumental role in establishing not only the U.S. Geological Survey but the Bureau of American Ethnography as well—headquarters for a nationwide study and documentation of the native cultures of what was now the United States.

While Powell respected and understood native cultures far better than almost any academic or government agent of his day, he knew that settlement was inevitable. While he studied the Indians to try and record what was being lost, he developed a keen interest in irrigation and carefully planned development from the Mormons as well. Unlike others who saw the West as a land of unlimited opportunity, Powell saw it as a land of limited possibilities, one where water was in short supply—an obstacle that overshadowed every other. His *Report on the Arid Lands of the West*, published in 1880, was a landmark work, detailing the region's chronic shortages of both usable water and arable land, outlining problems that still plague the region today.

In the meantime, work on the geology of the Colorado Plateau continued. Powell became head of the U.S. Geological Survey, but bureaucratic responsibilities along with an increasing interest in both Indian affairs and irrigation soon forced him to delegate much of the research he had planned to others. While he busied himself with political intrigues in Washington, he sent Grove Karl Gilbert to the Henry Mountains and the Great Salt Lake and Clarence Dutton to the Grand Canyon and the High Plateaus. After his early preference for friends and relatives, in his later years Powell proved as adept at picking personnel as projects.

Gilbert, Dutton, and Powell would reshape the way geologists thought about the world around them, but they had only the barest framework of ideas to guide them. The Grand Canyon, it was suggested, had been formed by the collapse of an underground river. The intricately carved desert landscape around it had taken shape during the Great Flood described in the Bible. As Powell and his coworkers traveled around the plateau, they mapped its major features and named its major rock units, developing conventions in geologic mapping and the naming of rock units still in use today, tying the names of rock units to regions where they were most clearly displayed—the Bright Angel Shale for its classic exposures along upper reaches of Bright Angel Creek in the Grand Canyon, the Mesaverde Formation for its appearance in the high cliffs of Mesa Verde in southwestern Colorado.

Geology as a science was still taking shape. Little more than fifty years old, most of its leading theories and ideas had been developed in Europe, where neither the features of the land nor its history bore anything but the vaguest resemblance to those of the American West. Studying the patterns and arrangements of the rocks around them they came up with new theories of mountain building and volcanism to explain the origins of the plateaus and peaks they saw in the deserts around them.

More than anything else, however, what impressed them was how much was missing. While the edges of the plateau country were capped by layers of relatively young rocks like those of the Mancos Shale and Mesaverde Formation, over much of its vast center those same rocks were almost entirely gone, eroded away in an event they called the *Great Denudation*. Built up in layers, the rocks of the Colorado Plateau had been stripped away in irregular pieces, carried away by rivers and streams. Those river and stream drainages, they soon realized, were not static but had a history all their own—cutting canyons, changing direction, alternately appearing and disappearing as the land rose and fell around them. To explain it they devised a series of categories and terms: antecedent, consequent, and superimposed drainages. In places they discovered even older erosional surfaces—pediments at the base of mountains and high plateaus that had been all but concealed by younger blankets of sediment from more recent episodes of erosion.

Studying the region's canyons, they theorized that its rivers and streams had cut rapidly downward until they reached some stable base level. Afterward, surrounding mesas and cliffs slowly followed suit, gradually wearing away to the level of the rivers below. There was an orderly progression to this wearing away as well: Once established, the lines of cliffs tended to perpetuate themselves as they retreated, preserving a particular trend or orientation. In places they seemed to have retreated by hundreds of miles. Differences in hardness between their successive layers of flat-lying rocks caused them to erode at different rates. While hard layers tended to form sheer, high cliffs, the softer units tended to form broad, open slopes, giving the walls of canyons and plateaus a distinctive stairstep look. In places, these softer layers of rock had eroded so quickly that they actually undermined harder layers of rock above them, causing them to break free from the faces of cliffs and tumble into the deserts below like icebergs calving off the face of a glacier.

Most puzzling of all, however, was the fact that rivers and streams on the plateau often seemed to cut right through its peculiar uplifts and plateaus—flowing right through them as if they weren't even there. In the case of the Grand Canyon, the Colorado River cut through the heart of the Kaibab Uplift, separating the Coconino and Kaibab plateaus, even though low, flat ground lay to either side. Other rivers managed the same feat: the Green, San Juan, and San Rafael. In places others like the Escalante and the Virgin cut through the rock in winding slot canyons whose walls were sometimes barely more than an arm's width apart—meandering through several hundred feet of solid rock like streams flowing across a flat sandy plain. What had happened, they theorized, was that streams had been in place long before the land had started to rise, meandering across a flat and almost featureless landscape. As the ground rose up beneath them, they cut down through it, carving their winding paths right into the rock. That canyon cutting, they theorized, had all taken place some 50 to 70 million years ago.

The broad framework of ideas laid down by Powell, Gilbert, and Dutton regarding the geological evolution of the Colorado Plateau are still in place, in many cases as valued today as they were more than a century ago. But while time and study have not changed many of their basic ideas, geologists today have a far different sense of timing. The canyon cutting that carved the surface of the Colorado Plateau did not take place 50 million years ago but perhaps as recently as 5 million years ago. While the great faults and folds that mark its surface today—features like the Defiance and Zuni upwarps and the San Rafael Swell—appeared some 70 to 50 million years ago during the Laramide orogeny that built the Rocky Mountains, geologists now believe that erosion may have flattened that ancient landscape out, filling valleys and lowlands with eroded rock and sand, burying the folds and uplifts of Laramide time under a blanket of sediment—to create a surface as flat and level as that of the even older deserts and seafloors that preceded it. Afterward rivers and streams began to migrate across it, setting the stage for the winding canyons to follow. As the land rose up not only did the rivers cut down through the rocks below but erosion stripped away those that blanketed the ground above as well, revealing the details of ancient faults and folds. Beyond the narrow reach of its canyons, this slow wearing away of the landscape has created the illusion that rivers here pay no attention to gravity or topography. Like low walls and fences hidden by drifts of snow, features like the Waterpocket Fold, the San Rafael Swell, and the Comb Ridge Anticline appeared, their paths cut by rivers and streams that had once flowed above them. Rather than avoid tilted piles and plateaus of rock, rivers here seemed to flow right through them. It is an illusion that is repeated over and over again on the Colorado Plateau.

Plate tectonics made it easier to picture the movements of continents and the rise and fall of mountain ranges, but for some reason it seems more difficult to picture the shifting patterns of deserts and forests; or the changing paths of rivers, lakes, and streams. It is not just the position of the continents that has changed but everything about them. After the close of the Laramide orogeny in Arizona, land to the south of what is now the Mogollon Rim was not a lowland but a highland whose slopes gave rise to rivers and streams that carried sand and sediment northward across the surface of what is now the Colorado Plateau.

The region's rise would eventually stand those streams on their head. In northern Arizona the Kaibab Arch appeared, a broad zone of uplift that includes both the Kaibab and Coconino plateaus that surround the Grand Canyon. Streams slowly began flowing not north but south, stripping away the overlying layers of younger rock, leaving the late Paleozoic rocks of the Kaibab Limestone, the 250-million-year-old rocks that line the rim of the Grand Canyon today, exposed at the surface. Meanwhile the canyon itself was nowhere to be found. Farther north on the Colorado Plateau the ancestral Colorado River flowed south from its headwaters in the Rockies. Its course, however, bore little relationship to the one it follows today. Instead of heading southwest through the rapidly emerging Kaibab Upwarp, the river headed southeast into the Painted Desert, emptying into Lake Biadochi, an ancient lake the same size and shape of Lake Erie that once reached from Flagstaff to Winslow, right through the heart of what is now a barren desert.

Around five million years ago, however, streams cutting headward through the Kaibab Upwarp eventually reached the ancestral river and "captured" its flow, sending the river southwestward through the heart of the Kaibab Uplift. While it took more than a billion years to build up the layers of rock that make up the walls of the Grand Canyon, it may have taken no more than a million years for the river to cut through them. Instead of flowing southeast into the Painted Desert, the river headed southwest toward the rapidly expanding Gulf of California.

Where the river went before it reached this new sea is uncertain. Some have suggested that it followed the course of the Little Colorado River, heading upstream toward Mount Baldy on the Mogollon Rim. Others believe that the river may have headed due west across the high plateaus of central Utah, through the area around the Uinkaret and Shivwits plateaus near the Utah-Arizona line. Then too, it may have simply evaporated in the desert reaches of Lake Biadochi without ever reaching the sea, like the playa lakes that lie scattered across the deep valleys of the Great Basin today.

Uplift and stream migration may have shaped the upper reaches of the Colorado River as well. Farther north near the Uncompahgre Plateau between Gunnison and Grand Junction, Colorado, is Unaweep Canyon, a deep wide canyon traversed by only a small, insignificant stream. The canyon itself, however, was apparently carved by the ancestral Colorado River. When the land began to rise, the river could not keep up and was captured and redirected to its current course through the Grand Valley toward Grand Junction in Colorado and Moab farther south in Utah. The canyon was simply left behind—the way a broad, eastern river like the Ohio or Mississippi periodically jumps channels as it seeks a shorter route to the sea, leaving towns and rail lines high and dry.

While movements of plates and the slip and slide of the continent along deep faults and ancient mountain ranges had defined the borders of the Colorado Plateau region, its surface would ultimately be shaped by nothing more complex than gravity and running water. The erosion that followed the Colorado Plateau's dramatic uplift in the late Cenozoic not only exposed ancient layers of sandstone and rock but also released them from the weight of several thousand feet of rock. Several hundred million years of uninterrupted deposition and sand had buried the rocks of the plateau under so much rock and sand that they were compressed. Released from this weight, their surface began to crack and break almost like a cold glass suddenly filled with boiling water. Walking across the slickrock at Arches or Canyonlands National Parks, you can see the effects of this unloading in a series of regular cracks and joints that run through the rocks in neat, straight lines, almost like a grid. Collecting points for soil and water, they appear in places as thin green lines on the bare rock—supporting tough desert shrubs and twisted trees: narrow lines of pinyon, juniper, cliff rose, and mountain mahogany. You can see the same features high above the canyon rims at places like Zion National Park and at Capitol Reef as well.

While the rocks were being unloaded from above, the center of the Colorado Plateau, the canyon country that borders the junction of the Colorado and Green Rivers near Moab, Utah, was being pushed upward from below just as the layers of ancient salt discussed earlier began to work their way to the surface. In places like the Needles district of Canyonlands National Park you can see how this uplifting has broken the land into fins and narrow walls of rock—slabs of rock ideal for the formation of arches and alcoves.

Arches form when the rock "spalls off" or falls off in slabs. Like cracks and joints, much of this spalling, as it is called, is due to the expansion of the rocks that followed their unloading. Elsewhere, however, the fracturing is aided by water that seeps into cracks and then freezes in winter. As the water turns to ice it expands, splitting the rock like a chisel or wedge. Forces like these helped create the alcove caves at places like Mesa Verde and Canyon de Chelly where the Anasazi built their picturesque cliff dwellings. Where the push of salt domes and the pull of unloading had broken the rocks into fins and narrow walls, however, this spalling created arches—bridges of solid rock arching across the desert like a rainbow. Elsewhere desert streams and rivers punched through these narrow rock walls to create natural bridges as well.

The smooth surfaces of the slickrock desert here give one the impression that rocks have been carved by the wind. Run your hand over an orange wall of Entrada Sandstone and you will find that it nearly crumbles beneath your hand. Most of the sandstones here are held together by a water-soluble cement of calcium carbonate. Rainwater seeping through the surface of the rocks slowly washes them away. The debris is then blown away and smoothed by the wind. Here and there the surface of the rocks are peppered with clusters of holes and pockets like Swiss cheese—a series of tiny rounded pockets and alcoves carved by the wind. Geologists call these pockets *taffoni.* No one is certain how they first formed, but it is thought that they may be due to some irregularity in the composition of the sandstones. Elsewhere sheets of purple-and-brown desert-varnish stain the surface of the rocks like spilled paint. The coating comes from small particles of iron and manganese. Rare in the rocks themselves, it is thought they may come from mist or rain and then are left on the rocks as the water evaporates.

Differences in hardness control the steplike retreat of the cliffs as well. Where a hard caprock covers the softer layers of rock below, solitary pillars and spires are temporarily left behind like sentinels. In places like Monument Valley, surrounding cliffs seem to have almost completely eroded away, leaving projecting piles of rock standing like giant statues in the desert—a record of ancient cliffs and rocks. In time erosion and settling will wear the Colorado Plateau flat, washing its canyons and plateaus away. Other disturbances may break it up further, or send its surface shooting up or down. The only thing certain is that the landscape we see today is fleeting and quickly disappearing—a brief instant in time that will never come again.

TWENTY·THREE

Rivers

ATE MORNING is the coolest time of day. At sunrise the air in the canyon was still and stale. The night before the temperature had not dropped below ninety degrees. At midnight the boulders and stones surrounding our campsite were still warm to the touch, like the side of an oven that has not yet begun to cool. By ten o'clock in the morning the desert above was already hot and bright. Down here in the canyon, however, the sun would not rise until almost noon. Looking up I could see the light inching its way down the canyon walls. As the rocks above began to warm, currents of air began to play through the canyon, bringing cool, fresh breezes. The liquid song of a canyon wren floats down from the smooth walls of sandstone that rise up above my head. By late afternoon the temperature will climb above 105 degrees.

Halfway up the face of the cliff in a high alcove cut into the curving red rocks is a small island of green—a hanging garden of flowers and trees. Picking our way up a slope of jumbled boulders and stones, we come out into a narrow forest of redbuds and oaks broken by stands of tall grass and reeds. Alongside the cliff clumps of flowering columbine, moss, and maidenhair ferns seem to grow right out of the rocks. Water percolates down through the porous layers of red and white sandstone above, until it meets a less permeable pink-

and-purple layer of shale below and comes trickling out of the rock—a seep in the canyon wall. There has been no rain for more than four months, but standing along the edge of the cliff you can hear the steady drip of the water. A few feet away the ground is blanketed with cinders and ash.

The alcove forest and hanging garden here is not wild or remote. It sits on the edge of Lake Powell in southeastern Utah in a side canyon a few miles from Hite Marina. For the past three days I have been camped out here on a narrow arm of the lake with Tim Graham, a biologist with the National Biological Survey studying the recovery from a burn. Five years before on a July 4th weekend this alcove forest was literally burned to the ground by a group of house-boaters celebrating the holiday with a barrage of illegal fireworks. With nearly two hundred miles of lake stretching out behind them they decided to set their fireworks off—not out over the water but up into the small oasis of flowers and trees clinging to the rocks above them. In the tinder-dry weather that precedes the arrival of the late-summer thunderstorm season here, it didn't take long for this small forest to burst into flames. Other houseboaters in the canyon video-taped the fire. In it you can see trees burst into flames like torches while flowers and ferns clinging to the rock walls behind them are transformed into balls of fire. The heat of the blaze was so intense that sheets of rock from the alcove shattered into pieces and fell to the ground below.

In five years the recovery has been mixed. While thickets of redbud have sprung from the stumps of charred trees and clumps of flowers and ferns have begun to spread across the cliff, the rich garden of plants that once spread out from the base of the cliffs to all but completely cover the floor of the alcove is nowhere to be found. Instead of flowers the ground is covered with cinders and ash and broken chips of rock. No plants, or even mosses and lichens, cover the ground here. In the dry desert air the ashes are so well preserved that they seem like the cinders of a fire that burned only a few days before. As we walk across the ground, our footsteps stir up waist-high clouds of ash. With the plant cover gone the ground here is completely dry, baked by the sun each day as it drifts through the sky overhead. When grasses, ferns, and mosses covered the ground here, moisture was wicked away from the back wall of the alcove all the way to the edge of the cliff. While trees and ferns growing near the seep have been able to make a startling recovery, the ground here in all probability will remain bare for decades.

Tim Graham grew up in Santa Fe in the 1960s, long before it became known as "the City Different," a mecca for tourists and well-heeled retirees. In college he was originally interested in studying marine biology, but he found the coasts too crowded with people for his liking. For the past few years he has worked as a biologist at Canyonlands National Park in Moab, Utah, a few hours' drive away. Transferred this year to the University of California at Davis and the newly created National Biological Survey, he is happy to be out of the city and back in the desert. Lake Powell, however, is far from deserted. From Glen Canyon Dam the water reaches northward for nearly two hundred miles— backing up not only the Colorado River but portions of the San Juan, Escalante, and Dirty Devil as well. Several hundred feet beneath its smooth, blue surface lies the Glen Canyon, once thought to be the prettiest reach of the Colorado River: not a deep, overpowering chasm like the Grand Canyon farther south, with its jagged walls of rock and all but impassable rapids, but a smooth-sided canyon of alcoves, grottoes, and hidden glens carved into soft layers of Navajo, Entrada, and Kayenta sandstone where the river ran quiet and deep.

Today small hanging gardens like these near what was once the canyon rim are all that remain of those riverside gardens and glens. The Glen Canyon National Recreation Area is flooded with nearly 3 million visitors each year—on par with even the crowded Grand Canyon farther south. They do not come to savor the desert but to play in the water on expensive, motorized toys—jet skis, speed-boats, and houseboats. There is no solitude here on the edge of the lake in midsummer: from morning twilight until well after sunset the air is filled with the steady whine of outboard motors and the rumble of inboard/outboard exhausts. The sound carries for miles, echoed and amplified by the canyon walls. At times I feel like I am standing on the edge of a rush-hour freeway in southern California or alongside the tracks of a busy subway station in New York City. From time to time a speedboat or jet ski peels off the main channel of the lake to explore the narrow side canyon where we work among the rocks. As it arcs and glides through the narrow arm of the lake below, the noise is almost deafening.

The shoreline here is often fringed by walls and shelves of bare rock. Here and there, however, are scattered beaches, narrow pockets of well-used sand littered with piles of trash and excrement that stink and swirl with flies. Those who travel on the water here do not travel gently on the land that surrounds it. The desert here is a landscape to be used, a place to dump one's trash, at best a spot to tie up for a few moments' rest before speeding off into the water again. Walk behind the ring of trash that fringes the lake, however, and you will quickly find yourself in the desert, wandering up narrow side canyons and washes whose floors hide groves of grass and cottonwood; or climbing out onto the smooth, sloping surface of the slickrock deserts above where the shapes of distant plateaus and peaks shimmer in the smoky heat.

Reaching the lake after a day's walk through the desert that surrounds it is a shock. There is something incongruous and startling about the sight of so much water in the midst of such a dry landscape. For the past three days the temperature here has climbed above 105 degrees. By late afternoon the heat is almost killing. We work as long as the shade lasts, starting each day at five in the morning, more than an hour and a half before the sun. We spend the afternoons looking for shade in the lee of rocks and cliffs, catching up on notes and reading—or simply trying to sleep. When the heat is at its worst, I jump into the lake and swim, staying in the cool water until my skin begins to pucker and curl. Graham, however, refuses to come in, but hunts for meager shade and continues to work and read. "I'm a desert rat," he says. "I don't swim," but we both know that is a convenient excuse. He is still thinking of the deep, smooth-sided canyon that used to cut through the desert here and the river that once ran through it. The water that now covers it he refuses to enter.

Nine years after his pioneering trip through the canyons of the Green and Colorado, John Wesley Powell released his *Report on the Lands of the Region of the United States, with a More Detailed Account of the Lands of Utah*. While its title was dry and academic, the report itself was nothing short of revolutionary. In graphic detail it pointed out the inescapable barriers to settlement and development that lay west of the 100th meridian—that vast dry reach of the United States that stretches all the way from the High Plains on the eastern edge of the Rockies to the Coast Ranges that border the Pacific—a place where rainfall is seldom more than twenty inches per year and often less, the bare minimum at which agriculture as it is practiced in the richer and more fertile East is possible. Settlement and development of the region, he argued, would have to be far differently planned and organized than elsewhere in the country if it was to have any chance of success. What was needed was nothing less than a new way of farming and a new system of land tenure.

Powell's trips through the Southwest had taken him not only across the Colorado Plateau but the Great Plains, Rockies, and Great Basin. He understood not only the region's geology far better than almost anyone of his time but its biology as well. Water was the key to life here—not so much the rain that fell from the sky but the water contained in the rivers and streams that ran across the ground, gathering the scarce rainfall from several hundred or even several thousand square miles of desert. That simple fact made life here far different than almost anywhere else in the United States. Ultimately it affected not only the distribution of plants and animals but of people as well.

Early maps of the American West often called the high reach of the Great Plains that lay west of the 100th meridian the Great American Desert. The land was unquestionably dry, but the appearance of that desert, as Wallace Stegner pointed out in his biography of Powell, *Beyond the Hundredth Meridian*, depended greatly upon the time of one's arrival. In the spring of a wet year it could seem like a vast green garden, with waving fields of grass and flower. By late summer it was often reduced to nothing more than a sun-baked plain. Settlement of the region after the close of the Civil War, however, would coincide with a steady but temporary increase in rainfall. While steel plows turned over the rich, virgin topsoil of the prairie, rain transformed the once dry land into phenomenally productive fields that yielded bumper crops of corn, wheat, and barley. Among farmers, speculators, and scientists alike, it became an accepted axiom that rain followed the plow. One had to merely till the soil and plant a crop and the rain would magically follow, turning the dry grasslands of Kansas, Oklahoma, Nebraska, and eastern Colorado into rich and fertile farmland.

To those inclined to see the westward push of the nation in divine terms it was yet another sign of heavenly favor, another confirmation of Manifest Destiny. With hard work and luck pioneers from Europe and the eastern United States could settle here and thrive, gaining title to farms of 40, 80, and 160 acres by investing nothing more than sweat and labor. For immigrants from Europe— England, Ireland, Germany, and France—those who had grown up under the heel of the aristocracy and landed gentry, it was like a dream come true: free land, a chance to finally own the ground beneath their feet. Never mind that it had only recently belonged to scattered bands of native peoples. Their lives and dreams were nothing like the immigrants' own.

While the mountains of the Rockies and Sierras spawned gold rushes, the plains and valleys beyond them spawned land rushes. Farmers and ranchers eager to stake out a claim headed west, where a variety of government programs were doling out land to those willing to settle it and make improvements: the Homesteading Act, the Desert Lands Act, the Timber and Stone Act. Rules and regulations encouraged and directed development, requiring would-be owners to plant fields and windbreaks of trees in the case of farmland. The size of the family farm was set at 160 acres.

At first would-be settlers leaped right over the Colorado Plateau and the arid lands that surrounded it, crossing the Great Plains and heading through the mountains and deserts to California, Oregon, and Washington. After the close of the Civil War that tide of immigrants would begin to break and roll back eastward again, into Utah, Colorado, New Mexico, and Arizona—the wide dry reach of the Intermountain West. The success of Mormon farmers gave others a suggestion of what was possible here, but it was one thing to settle along the well-watered reach of the Wasatch Front and the High Plateaus and quite another to venture out into the deserts beyond. If settlers here were to have any hope of success, Powell believed, things would have to be done far differently.

While others spoke of rain following the plow and irrigation making the desert bloom like a rose, Powell spoke of limits and deficiencies. While others claimed that the region could support a popu-

lation of several hundred million, Powell suggested that in all probability it would be lucky to support more than a few million. While
the land was vast, less than 20 percent of it was arable, he said, and
in the case of Utah that percentage was no more than three. Even
then, he added, water was in short supply. Its use would have to be
carefully partitioned and planned if settlers were to have any hope *study*
of success. Rivers, streams, and springs were critical. Those who
controlled the water would control the land. A claim of five or ten
acres along the banks of a river could in effect tie up thousands of
drier acres beyond. Rather than dividing the land up into straight-
sided squares and rectangles, as surveys had done in the East, it
made far more sense to organize the region around rivers, basins,
and drainages and all the irregularities of shape and topography that
implied. If the land and its water were not carefully divided, he
warned, a handful of speculators and corporations could well end up
controlling it all. Finally, he suggested that the size of homesteads
should be changed as well. Irrigated farms were phenomenally productive, while dry land farms in the region could produce only a
fraction of those farther east. For irrigated farms he proposed cutting the homestead size in half—from 160 acres to just 80; dry land
farms and ranches, he suggested, should be increased to a sprawling
2,560—four square miles of land. Given the spareness of the land
here, it was the minimum a family would need to survive.

In an era when most public debate was governed by persistent
optimism and dominated by an almost mindless euphoria for the possibilities of growth and development, Powell's views were considered almost heretical. In Congress, senators and representatives
from the West derided Powell as an easterner meddling in Western
affairs, a "collegiate fledgling" and a "scientific lobbyist" intent on
building his own power base—a contempt for scientific study and
impartial analysis that western congressman have long since per-
fected to an art form. To be sure, Powell's proposal would have given
him a great deal of power, but his intentions were far more
respectable than those who criticized him, and his assessments were
far more realistic. Even so his proposals were ridiculed by the press
and eventually dismissed by Congress.

Over the next ten years, however, the booming economy of the
West would collapse. During the winter of 1886–87, a severe bliz-

zard would all but destroy the region's cattle industry, ruining both fortunes and families. Those who managed to survive the winter were soon hit by a prolonged drought that lasted for nearly five years. Powell had warned that the region was prone to periodic droughts—and now his doubters were getting a quick and brutal education. Crops and cattle died in the fields. Farms and ranches were abandoned by the thousands. Meanwhile back east, the Johnstown Flood in Pennsylvania killed more than two thousand as a poorly designed privately built dam on the South Fork of the Conemaugh River collapsed. The towns and villages scattered across the valley floor were swept away in a wall of solid water.

In 1888 many of the same western senators who had mocked Powell's ideas in the past suddenly brought them back to life. Now head of the U.S. Geological Survey and the Bureau of Ethnology, he was put in charge of a National Irrigation Survey and given the power and funding he needed to bring his ideas to life. While Powell sent his survey teams to the West to map the location of farmland suitable for irrigation and the possible sites of dams and reservoirs, the entire region from the High Plains to California was closed to settlement and sale. Powell was methodical. The work would take not months but years, and he made no apologies. He thought it would be criminal to allow people to establish themselves on ranches and farms where they had no chance of survival. Carefully planned irrigation projects would assure them a decent chance of success, creating a small but well-organized society of family farms and ranches. If the West was going to be settled at all it needed to be settled carefully and logically.

The politicians who supported Powell, however, had not bargained for such a long delay, or perhaps they finally realized just how much power they had given him. Land speculators began following Powell's surveyors around, offering lots alongside the newly mapped reservoir sites for instant sale. Others complained that the region had been locked up—with nothing more than a pile of maps to show for it. There was, they argued, something profoundly un-American about it all. By 1889 Powell would find himself testifying before congressional fact-finding committees and pilloried in the press, fighting to keep both the Irrigation Survey and the U.S. Geological Survey alive. Powell's estimates, they argued, were too conservative, his assessments too pessimistic. They lacked the grand dimension of western life, the sense of limitless possibilities. True, the land might prove to be not all it was hoped to be—some might even fail. But that was acceptable too. There was money to be made in failed farms and ranches; money to be made in the sale of seeds, supplies, and land; money to be made by those who held the mortgages and the loans—empires that spread across thousands of acres to be built up from a quilt of failed farms and ranches. For banks, monopolies, and holding companies, failure could be as profitable as success.

Political maneuvering and manipulation would eventually shatter Powell's plans for a carefully planned settlement of the West, and the results would be largely as he had predicted. Of the nearly one million families who sought homesteads west of the 100th meridian, more than half would fail. In places failures and foreclosures were as cyclical and predictable as drought, as towns and farms collapsed and failed and were then revived—only to fail again.

By default the government would eventually come to direct irrigation in the West. The Newlands Reclamation Act that began the slow but steady process of bringing water to the West was finally passed in 1903, the year Powell died. In less than fifty years, the Bureau of Reclamation would lace the West with reservoirs and irrigation canals, creating green fields and orchards in the deserts of California and Arizona and neighboring states as well. That progress, however, was accompanied by a steadily growing number of problems.

While the large reservoirs that line the Colorado—Lake Mead, Lake Powell, Flaming Gorge, and others—hold enough water to last the region's cities, farms, and ranches several years, every drop of it is spoken for, often several times over. In the 1930s when the Colorado was apportioned between the states through which the river and its tributaries run, its yearly flow was estimated at 17.5 million acre-feet—an acre-foot being the amount of water needed to cover an acre with a foot of water. The problem is that the engineers based their calculations on data gathered from a few years of what proved to be abnormally high rainfall. While the river may be apportioned at 17.5 million acre-feet, its average yearly flows are in fact no more than 11.7 million acre-feet. The shortage has created a welter of problems that affect us still today—constant legal sparrings and courtroom battles fought by competing armies of well-paid attorneys and advisers. Stored behind dams in vast desert reservoirs, millions of gallons of water vanish into thin air each day due to evaporation. What remains grows steadily saltier. Elsewhere irrigation of poorly selected mineral and salt-laden soils makes the problem even worse as runoff carries nearly toxic levels of salt and minerals like selenium back to the rivers and reservoirs. By the time the Colorado River finally reaches Mexico, its waters are saltier than the sea—unfit to drink or even water crops. And evaporation creates problems in irrigated fields as well. As the heat of the sun draws off moisture, salts and minerals are left behind. Today the United States is losing more irrigated farmland to salt buildup than it is bringing into production with new irrigation.

While the feats of engineering behind the region's dams are remarkable, their economics are questionable at best. Billions of dollars of investment have brought an area roughly equal to the size of Missouri into production. The reservoirs, dams, and canals that made it all possible were envisioned as long-term loans. Few irrigation districts, however, have been able to keep up with their scheduled payments—even when the "loans" were interest free. As for the day-to-day costs of the projects, water users pay only the costs of delivering the water to them. The water itself is free. The balance has been picked up by taxpayers. No less important to western water users than the canals that bring them water is the pipeline of public money that keeps the system running.

In California the water from publicly funded water projects goes to water the fields of "farmers" like Exxon, Tenneco, and Getty Oil—sprawling corporate farms that cover thousands of acres. Pioneers like Powell saw irrigation as a way of settling the West and building a stable community of small, independent farmers. In California, however, politicians have preferred to promote not agriculture but agribusiness, tailoring the state's water system to meet the needs of wealthy corporations. They lobbied for changes to federal laws as well—raising the acreage limits for recipients of federally sponsored projects from 160 acres to 960 acres, giving corporate farmers access to federally funded water projects as well. Corporations, they argued, could do more for the state's economy than family farms. Instead of farmers working their fields, the crops here would be tended by seasonal field hands.

In spite of their profitable orchards of oranges and almonds and their productive fields of lettuce and grapes, if California growers were forced to pay the full costs of the irrigation projects that bring their water, they would find themselves running in the red. Instead of a means of promoting self-sufficiency, over much of the West irrigation has become an elaborate form of public assistance, of welfare for the rich—a sink for hard-earned tax dollars. Elsewhere the economic figures are even worse, in some cases balancing out to subsidies of several hundred thousand dollars per farm. In higher, colder states like Idaho, Wyoming, Colorado, and Utah, farmers put their subsidized irrigation water to work raising crops like cotton, corn, wheat, and barley that bring even lower prices—crops so plentiful elsewhere in the country that farmers in wetter, more fertile eastern states like Arkansas, Missouri, Illinois, and Iowa are actually paid not to grow them—all part of a federal program to keep the market prices for these basic crops from collapsing on an already flooded market.

One wonders what Powell, a champion of practical and careful development, would make of the lake that now bears his name. Lake Powell stretches across the desert for nearly two hundred miles, and yet it provides not a single drop of water for irrigation. Glen Canyon Dam at the head of the lake was built to provide cash, not water. The income generated from the sale of electricity created by the water that streams through its turbines is used to cover the costs of irrigation projects whose shaky economics made it impossible for them to ever stand alone.

Dams and reservoirs changed everything about the rivers of the Colorado Plateau. When John Wesley Powell ran the Colorado River in 1869, its waters were so laden with silt and sand that the water was actually red—the effects of canyon cutting and the steady wearing away of brightly colored layers of rock in the deserts that fringe its reach. Today the water that pours out of the base of the Glen Canyon Dam is as clear and cold as the mountain streams of Colorado and Wyoming where the river first begins. Blocked by the dam, the river is transformed into a still, deep lake. As the water slows and settles, its heavy load of sand and silt drops below. Instead of cutting a canyon here, the river is filling one in. In little more than five hundred or six hundred years both the ancient canyon and the reservoir that now covers it will be completely filled with silt.

Downstream, changes in the river affected not only populations of fish in the water but plants and animals alongside it as well—changing not only the temperature and clarity of the water but also the patterns of flows and floods. Up and down the river groves of cottonwood trees gave way to thickets of tamarisk. Native fish like humpbacked chub and Colorado squawfish were replaced by trout and pike. In the 1960s when the dam first went into operation, operators turned the river on and off like a light, opening its gates during the day to send water speeding through its turbines when power demands were high and the price of electricity was at its peak and closing them off at night. Over the course of a day river flows could vary anywhere from ten thousand cubic feet per second during the day to as little as a thousand cubic feet per second at night. Rafters running the river below the dam often found themselves becalmed, the flow of the river in the upper reaches of the Grand Canyon reduced to a trickle, forcing them to wait for the powers that be upstream to turn the river back on and allow them to continue their ride. Later these plans were changed to even the river out, leaving flows of a steady few thousand feet per second all day.

Soon, however, rafters began to notice that the water was beginning to strip away bars and beaches along the river—broad reaches of sand where they once stopped to camp and rest. The carefully regulated flows were unlike anything ever seen on the river before. In predam times spring floods might find the river within the Grand Canyon running at more than one hundred thousand cubic feet per second while summer flows sometimes reached no higher than a few thousand cubic feet per second. The dam, or so the common wisdom ran, had stripped the river of sand. The clear, cold water that poured out of Glen Canyon Dam was sediment-starved, hungry water that replenished itself with sand and silt from the riverbed as it headed south toward the Grand Canyon and the deserts beyond. Now, however, many are beginning to believe that the origins of the problem are more subtle and complex.

Scientists like the U.S. Geological Survey's David Rubin believe that the loss of sandbars and beaches along the river may actually have more to do with the loss of the river's annual floods than a loss of sediment. An expert on the movement of sand and sand dunes, for the past few years Rubin has been a member of a team of scientists studying the river as part the Glen Canyon Environmental Impact Statement. While the dam is already in place with a lake spreading out behind it, managers and scientists are hoping to find ways to regulate the flows of water through the dam to lessen the impact on the rivers that lie both above it and below it.

While the clear water that leaves the dam has stripped the first few miles of the river of sand as predicted, Rubin said, by the time it passes the mouth of the Paria River, the waters of the Colorado are once again nearly saturated with sand. In fact, he said, studies have found that there is actually more sand in the riverbed once one gets below the Paria today than there was before the dam was built— even though the river's bars and beaches are quickly disappearing.

At first glance those facts seem contradictory. In the past, Rubin explained, floods roared through the canyons each spring, moving vast amounts of both water and sand—even boulders and stones—down the river. As the water began to slow, however, layers of sand and silt were deposited, left behind as beaches and bars as the river continued to drop. During the low flows of summer and fall they were left high and dry, safely beyond the water's reach. The Glen Canyon Dam largely put an end to those floods—reducing the river's wild flows to those of a nearly steady year-round stream. With no spring floods there was no way to build up bars and beaches beyond the regular reach of the river. Regulated by the dam, the Colorado flowed through the Grand Canyon like water pouring through a pipe, smoothing the sand beneath the river into a flat, even layer, carrying away beaches and bars.

Changes in the way the dam is operated may help restore sand-bars and beaches along the Colorado River below Glen Canyon Dam, Rubin said—small, carefully planned spring floods that send a surge of water and sand down the river. The process, however, will not be easy. There are a number of things to balance: the demands of irrigators downstream, markets for electric power, the needs of fish and wildlife, the effect of flooding on fragile archaeological sites along-side the river, and above all the region's overallocated supply of water. By careful planning and studying, Rubin and others hope they can find a way to bring the river back to life.

The tops of the bluffs are low, perhaps no more than a hundred feet above the flat desert below. Fifty miles to the north the Uinta Mountains score the edge of the horizon. In late summer the crest of their high, rounded peaks are green and free of snow. Seventy miles to the south the flat, square lines of plateaus run through the edge of the sky—the East and West Tavaputs plateaus that rise above the Book Cliffs of central Utah, their tops carpeted with forests of pine and fir. Here in the heart of the Uinta Basin, however, is a barren desert of dry, golden grass, blackbrush, and sage. It stretches for miles in every direction, its gently rolling surface cut here and there by dry washes and low ledges of sandstone and shale. The cover of plants is so thin that you can see the rock and soil below—colored dark brown and tan—as if the landscape here had been burnt by the sun. In places low cliffs and benches of pink and purple shales rise up out of the ground. Here on top of the bluffs there is no soil at all. The ground beneath my feet is covered with a carpet of smooth, rounded stones—desert pavement—varnished with shades of purple and brown, a carpet of gravel and stones left by glaciers that flowed out of the mountains to the north a few thousand years ago.

Not far from the base of the bluffs the Green River curls and winds through the desert below. After slicing through the center of Split Mountain to the north, the river runs through the flat, open deserts of the Uinta Valley for nearly seventy miles toward Desolation Canyon and the high plateaus to the south. The river here, however, does not lie at the bottom of a deep, narrow canyon but runs through the midst of a wide, shallow floodplain—surrounded by a narrow ribbon of green: galleries of cottonwood trees, meadows, and shallow marshes filled with lilies and reeds. Here and there the wide, meandering turns of the river enclose bottomlands—some still covered with shallow pools of water, others with bright fields of grass. On the far side of the river a small herd of antelope grazes in the center of a wide, open reach of now dry river bottom. Farther north the surfaces of marshes and ponds alongside the riverbank are dotted with the bright white shapes of pelicans and darker flotillas of ducks and geese.

It is late August and I am traveling around Ouray National Wildlife Refuge with Dan Schaad of the U.S. Fish and Wildlife Service. For the past week the temperature here has climbed to ninety or a hundred each day. This morning, however, the sky is overcast and gray with the clouds of an unexpected late-summer rain. From time to time the rain seeps out of the sky like a fine mist or fog—a rare occurence here in the deserts of northwestern Utah. I had arrived at the refuge a few minutes after seven; after talking for a while in Schaad's office, we had gone out to drive around the dirt roads that run through the refuge in his government-issue four-wheel drive. We stopped from time to time to look over flocks of birds with binoculars and a spotting scope. In less than an hour we have seen perhaps more than forty different kinds of birds: great blue herons and white-faced ibis wading in the shallows of marshes and ponds; kingfishers and swallows circling over the river; sand-pipers and plovers probing sand flats and bars of mud; hawks perched in tall, dead trees; bitterns and golden-crowned night herons stalking amid thickets of rushes and reeds. Here and there were animals as well: muskrats, deer, and coyotes.

The river is a magnet for life in the desert. In the spring the ponds and backwaters here are alive with several thousand ducks. In the fall migrating flocks of sandhill cranes pass—a hundred or more at a time. Tall cottonwood trees along the river hold rookeries of great blue herons and double-crested cormorants. More than just a hanging garden amid the canyons or a hidden alcove filled with trees, the land here is green for several miles. The boundary between river and desert, however, is sharp and distinct. A few hundred yards from the riverbank the land is bare and dry.

Ouray National Wildlife Refuge lies some thirty miles south of Vernal, Utah, in the state's northeastern corner. For more than twelve miles the Green River loops and winds through the refuge, a broad shallow stream whose surface is marked not by rapids but sandbars and sand flats. Riparian areas like these are relatively rare on the Colorado Plateau. Rivers here often run in deep canyons, sur-rounded by high walls of rock. From time to time, however, the region's major rivers—the Colorado, Green, and San Juan—emerge from their canyons and head across open valleys and deserts. Quite often, however, their path is fringed with dikes and dams protecting farms and fields onshore—the rivers here are focal points not just for plants and animals but also people.

For a few miles here in northern Utah, however, the river comes into its own—a source of richness and life. The refuge was created in the 1960s to make up for riparian or streamside habitat lost due to the construction of the Flaming Gorge Dam more than a hundred miles upstream in Wyoming. Dikes along the river here hold water in, not out, creating a maze of shallow marshes and bottoms, mimicking the wet bottomlands that once characterized the open reaches of the river, providing a habitat for shorebirds and waterfowl and other birds and animals. In years when the river is high, Schaad said, they can simply open the floodgates along the dikes and allow the water to fill the bottomlands behind them. In dry years, however, they are forced to fill them with pumps along the river. "Historically," Schaad said, "you could probably buy up a parcel of land and just let it sit. Now, however, everything has been so altered that you've got to do some hands-on managing. This place is not an island."

What really keeps this area going, however, is the Yampa River, whose mouth lies only a few miles upstream. The Yampa, Schaad said, is still free-flowing, its path unmarked so far by any major dam. While dikes, ironically enough, protect the ponds and pools behind them, much of the rest of the area here still floods each spring, largely from the flows of the Yampa—a fact that helps keep the area here far more natural and wild than land either up or down the river. You can see it quite clearly in the thick groves of cottonwood trees that border the river. While they fill the air in early summer with millions of seeds that float through the air like fine tufts of cotton or wool, they depend on floodwaters and the thin layers of soil brought by floods to germinate and thrive. The forests here are healthy, a mix of young and old trees. Farther north along the river at places like Seedskadee Wildlife Refuge in Wyoming, however, old trees still line the river, but young ones are almost impossible to find. With the frequency of floods blocked by dams, the forests of cottonwoods that once lined the river are slowly dying off.

Even more important than the trees along the river or the birds that gather near the water are the fish that live in the river: endangered species of native fish like Colorado River squawfish and razorback suckers that the construction of reservoirs and dams on the Colorado Plateau have nearly pushed to extinction. The refuge here is one of the few places on the Colorado River system where these rare fish are still able to successfully reproduce. Rather than fight the river or focus on providing a habitat for ducks and geese, managers at the refuge are increasingly beginning to look for ways to work with the river, restoring its natural patterns and rhythms as much as possible, bringing back the seasonal floods and changes of flow that were once so critical.

Dams changed the upper reaches of the Green River just as the Glen Canyon Dam changed the lower reaches of the Colorado River farther south. Like the Colorado, the Green River had been a wildly fluctuating river, with flows often cresting at thirty thousand cubic feet per second during the spring snow melt and dropping to as little as five hundred cubic feet per second in midsummer. Dams changed all that, stabilizing flows over much of the river's reach to a few thousand cubic feet per second. Along with the dams new species of fish were added. Early Mormon pioneers had introduced catfish to the Colorado and the Green. Now others were added as well: game-fish like trout, pike, and muskie, fish that were not only adapted to cold water but were competitive and predatory as well. Dam construction was at its peak here in the 1950s and early 1960s. By 1966, the U.S. Fish and Wildlife Service was already tracking the demise of the river's native fish.

"It was a one-two punch," said Tim Moody, director of the Colorado River Fish Project in Vernal, Utah. The construction of dams and the introduction of nonnative species of fish all but wiped out the native fish. "It went from being a system with high variability and low diversity to a system to low variability and high diversity," he said: from an unstable river with only a few species of fish to a stable river with a great number of species of fish.

The native fish here were unique, found nowhere else but on the Colorado River and its tributaries; not high up in the rivers—the uppermost reaches of the Colorado, San Juan, and Green are trout streams—but down in depths of canyons and the floors of deserts to the south. During the Ice Ages when the climate was colder, they may have been forced far to the south. Plants and animals on land are not the only things that move with changing levels of tempera-ture and rainfall. Today there are more species of nonnative fish in the river than natives. The Green River as it spills out from the base of Flaming Gorge Dam is a world-class trout stream—another example, Moody said, of just how much the river has changed.

Today four of those native species of fish are listed as endan-gered and protected by the Endangered Species Act: the razorback sucker, the Colorado River squawfish, the humpbacked chub, and the bony-tailed chub. While the fish are all native to the Colorado River system, they have different habitats and needs. The hump-backed chub, for example, lives deep in the canyons in waters that are often fringed with rapids. Squawfish and suckers prefer broader, open reaches of the river—particularly for spawning. While little is known about the humpbacked chub and its close rela-tive, the bony-tailed chub, which may already be extinct, scientists have been able to study the other two in detail. In the case of the Colorado River squawfish and the razorback sucker, the short reach of the Green River that lies within Ouray National Wildlife Refuge may hold the key to their survival.

The fish depend on floodplains and backwaters to rear their young. Like salmon they migrate up and down the river to spawn, returning year after year to the same place. Unlike salmon, however, for squawfish and suckers it is not a one-way trip. Squawfish live to be more than sixty years old and spawn each year for decades as soon as they reach maturity. By attaching radio transmitters to fish, scientists like Moody have been able to track their migrations and movements. They seem to congregate on just two short reaches of the river, both upstream from the open waters and floodplains of the wildlife refuge. Suckers travel as much as eighty miles along the river to reach their native spawning ground; squawfish travel even farther—as far as two hundred miles. They lay their eggs in the spring when the river is nearly ready to crest. The eggs hatch in just four to six days. When the river crests and the water level begins to fall, the young fish or fry drift with the current. When the fish emerge they are incredibly small. Some biologists, Moody said, call them threads with eyes. How they know the point of their origin is not clearly understood. When the fish are young, their bodies con-tain high levels of a chemical known as thyroxine—the same chemical in salmon that is believed to be linked to their ability to find their native stream.

The short stretch of the Green River that runs through Ouray National Wildlife Refuge has the largest viable population of Colorado River squawfish in the Colorado River system—somewhere between two hundred and six hundred fish. While far greater numbers are found in Lake Mojave on the Colorado below Hoover Dam—between thirty and forty thousand—there are no young fish in the lake at all. Since the dam's construction their traditional spawning grounds are out of reach and quite possibly destroyed. While the population here on the Green River is small, it is still viable, Moody said, because the river's bottomlands and back channels still serve as a nursery for young fish.

The adults lay their eggs shortly before the river peaks, giving time for floods to fill in channels and backwaters below. Those pools of slow-moving protected water provide a habitat for young fish, a place to grow and mature away from the faster, more competitive reach of the river. Restoration programs, Moody said, will never restore the fish to their historical levels—but they can assure a stable population. "We have our floodplains," he said. "It's a winnable situation. We will never get them back like they were in the past, but we can get them back to sustainable numbers—and that is all the Endangered Species Act requires."

The Fish and Wildlife Service has been raising species of these fish in hatcheries for several years. How to proceed from here, however, is a subject of intense debate within the agency. Some argue that they should begin releasing fish to build up disappearing populations. Others like Moody argue that habitat is the key. "You can release all the fish you want to," he explained, "but if they don't have a place to live in there is no way they will ever survive."

Restoring and protecting habitat, however, is a far more tricky issue than simply raising and releasing fish. In the case of the Green River, it will mean careful cooperation with those living and working along the river, both above and below the critical reach of the wildlife refuge. While hydrologists and biologists working in the Grand Canyon have begun to release experimental "floods" from Glen Canyon Dam to rebuild beaches and sandbars in the Grand Canyon, biologists here have begun to experiment with sudden releases of water to recreate the floods that once filled the backwaters and side channels of the Green River in the Uinta Basin, providing a habitat for fish. The first test was disastrous—flooding the fields of ranchers and farmers farther upstream along the river. People were outraged, and rightly so, Moody said. In the future they will have to work more closely together if they are to have any chance of success. "If we increase flows we need to mitigate the damages to landowners. You have to make it a win-win situation," he said, "and convince locals they have a stake in it too—whether it comes in the form of payments and conservation easements for fish or simply keeping people informed and involved." Like anywhere else a few locals are vehemently opposed to the Endangered Species Act and the implications of the fish restoration program—but most, he said, are simply trying to make a living. Having promised flood control and a steady supply of irrigation water, and having encouraged them to build up farms and ranches along the river, federal agencies seeking to ensure the survival of endangered species of fish, Moody said, have to work with them to help keep them in business as well as bring the fish back.

While federal programs have had great success in restoring populations of dramatic animals like eagles and osprey, building up support for obscure species of fish is a much harder job. The fish, Moody said, are native to the rivers here and uniquely adapted to it. Whatever aesthetic or moral values one might attach to them, they are, he said, invaluable indicators for tracking the river's health. The West has water for farms and fields and suburban lawns, but it remains to be seen whether it has enough water or patience for species of rare and endangered fish. "We live in society and we have to support that society, but there are a lot of things we have been providing for people that aren't really necessary," Moody said. "What it really comes down to is values. What you value and what your neighbor values. And your neighbor may value his green lawn more than some humpbacked chub up in a canyon somewhere." The Colorado Plateau is no longer a wilderness. It remains to be seen how our modern society will adapt to it—whether we will allow the past some part of the future, making room here for others besides ourselves.

T W E N T Y · F O U R

Exiles and Refugees

DALE DAVIDSON IS AN ARCHAEOLOGIST with the Bureau of Land Management in Monticello, Utah. From time to time he gets unusual requests for information—calls from Sedona, Santa Fe, Boulder, and Seattle asking for advice on finding Fremont and Anasazi ruins. The callers are, they say, "really interested in Native American religion and New Age spirituality." Could he possibly recommend a site were they might go and "do a ceremony"? There are, of course, no sites suitable for such self-styled ceremonies, and Davidson offers them no encouragement, but they flock to the Four Corners area anyway—drawn by the ruins of ancient pueblo peoples, in search of visions in the desert.

Out in the field studying and surveying archaeological sites, Davidson finds regular signs of their presence: crystals hung inside of pueblo ruins or the remains of newly made pots shattered on the floor—all part of some elaborate and recently fabricated ritual. There are often other leavings outside as well: medicine wheels and spirit vortexes laid out in patterns of rocks and stones; or piles of ash and charcoal in front of pictographs and petroglyphs from "ceremonial" fires. They are not offerings left by Hopi, Zuni, Navajo, or Ute, but the debris of rituals of self-described New Age spiritualists. Most are Anglo and white, refugees from the suburban sprawl of

Denver, Phoenix, Los Angeles, and Seattle, in search of some link with the past or the land. While they view their crystals and medicine wheels as worship, tribal religious leaders often view their desperate attempts at imitation as a desecration, damaging and defiling their sacred sites and the ruins of their ancestors.

"They are messing around with things they don't really understand," one Zuni told me. "It would be better if they just left things alone." Other Navajo I came across deride them as wanna-bes. For the Navajo it is a simple matter to send unwanted visitors on their way when they show up uninvited at a sing or dance. For Pueblos, however, preserving the integrity of their ceremonies is a far more difficult problem. By tradition kachina dances are done for the benefit of the community, and portions of many ceremonies are both well publicized and open to the public. Guests are often welcome, but to preserve the power and integrity of the ceremonies, cameras, tape recorders, and sketchbooks have been banned for years during feast days in most pueblos. Even so, from time to time you will find a few outsiders lurking around the edges of dances and processions with carefully concealed cameras, and tape recorders trying to discover the details of rituals and chants. Fliers and advertisements in Flagstaff and Sedona as well as elsewhere around the region advertise quick courses in healing ceremonies and the making of prayer plumes and fetishes. There is something profoundly ignorant or arrogant in all of this—or perhaps a sad combination of both—as if the intricate details and discipline of native religions were something one could master in a weekend seminar or a few hours of careful reading; as if a few night classes or a correspondence course could train one to be a rabbi, priest, or minister.

For archaeologists the use of ancient sites and ruins for New Age rituals is merely yet another type of new and troubling damage. Make no mistake about it, the crystals, homemade pots, and fires have an impact on the ruins and remains of those they claim to worship: damaging fragile walls, staining panels of petroglyphs and pictographs with smoke while the flames cause rocks to split and crack—destroying fragments of art and culture that have lasted for centuries. While these signs of damage are growing, they are fortunately still not pervasive— merely another facet of the steady wearing away of the region's landscape and history brought on by a flood of tourists and visitors.

Since the turn of the century archaeologists estimate that something like 50 percent of the region's archaeological resources have been lost to development and theft. Oil fields, uranium mines, cattle grazing, and the growth of cities and towns have taken their toll on the region's archaeological sites. Pothunters have raided pueblo ruins and burial sites for precious pots and jewelry since the 1880s, when the Wetherill brothers' discovery of Mesa Verde put the region and its ancient cultures on the map. They hit both public and private land, digging up sites with backhoes and even bulldozers. A single pot can bring fifty thousand or sixty thousand dollars on the black market—sold to wealthy collectors in New York, Berlin, and Paris.

The handiwork of thieves often leaves fragile archaeological sites looking as if they had been hit by a bomb, in the words of one local archaeologist. But bad as it is, the destruction caused by pothunters has begun to pale under the steadily increasing impacts of the region's rapidly growing number of visitors. "The problem is people who love the place," said Ricky Lightfoot with the Crown Canyon Archaeological Center in Cortez, Colorado. "They're loving it to death."

Since the 1970s tourism in the Four Corners area has exploded. Visit the ruins at Mesa Verde, where more than seven hundred thousand visitors per year pass through, and you will hear not just English but German, French, Italian, and Spanish. With the park's restored and carefully managed ruins literally bursting at the seams, crowds of visitors have begun showing up at other less developed sites as well. Out on Cedar Mesa, a once remote concentration of ancient pueblo ruins not far from the Abajo Mountains in southeastern Utah, the number of visitors has gone from just four hundred to five hundred per year in the early 1970s to more than six thousand per year today, the BLM's Davidson said. While the rise in use is phenomenal, it has become all too typical of what is happening throughout the region.

"Recreational use is a big problem now," Davidson said. "That's not to say that other things like cattle grazing don't have an impact," he added. "But the problem with humans is that we're dealing with what people will be doing. The list of things a cow will do when it's left on its own is pretty finite. Versus that which people will do, which is infinite. Cattle don't want to go and see every archaeological site. They don't want to climb into every nook and cranny." Visitors in southeastern Utah come in all shapes and sizes—everything from "mom and pop in their RV to those who want to come and see the country in depth. It's public land and that's fair enough," he said. The problem lies in educating people as to just how fragile the region's archaeological resources are.

It's easy to tell a site that's had too many visitors because of the things that disappear, Davidson said. The first thing to go are the ceramics—the potsherds that litter the ground at many pueblo ruins, particularly the large, well-decorated ones: chips of red-and-black or black-on-white pottery. "People tend to collect these things despite whatever efforts we make at public education," he said. "People say, 'I'll just take one.' But if you've got ten thousand people visiting a site over the course of a year that's ten thousand sherds gone."

The problem, he explained, isn't a single, specific visitor, but the cumulative effects of several thousand. The land here is so spare and open that it seems almost untouched to many visitors. Finding an unmarked pueblo in the curve of a canyon or a panel of pictographs on the flank of a mesa leaves many with the illusion they are seeing it for the first time. Such ruins and artifacts are not hard to find. In San Juan County, Utah, alone, there are some twenty-five thousand documented archaeological sites on file at the Bureau of Land Management—and that may be no more than 10 percent of the area's total. Out of that phenomenal number, however, Davidson said archaeologists have found only one in recent years that may have been undiscovered since its abandonment several centuries before. And, he added, "We are quickly trying to figure out what to do with it before it does get discovered."

Trailing across sites is also a problem as visitation grows. Visitors walking across the floors of dry alcoves and middens leave a web of tracks behind. "They're just loose banks of material, and any kind of walking across them causes them to deteriorate really quickly," Davidson said. Some, he said, are drawn to the sites for photography, and they try to climb up to places on the walls for the best angle or a different perspective, damaging fragile roofs and walls. Others simply like to climb and crawl around simply for the fun of it. The BLM, Davidson said, has a few kivas open to visitation. While they have been shored up and partially rebuilt to make them safe, checking the site from time to time they've found the roof beams cracked by the weight of too many people gathering on the roofs. "You get this kind of damage in any kind of building," Davidson said. "And then you add to that the fact that these are very old buildings and you really get the kind of wear and tear you would expect."

Time, of course, takes its toll on ruins and remains. Pictograph panels fade and wear away. Pueblos and kivas tumble and fall. Those losses are unavoidable, Davidson said, and need to be accepted. But the impacts of tourists, visitors, and those just passing by are something else again. How best to control those impacts, however, is not by any means a simple problem.

"I still believe that you have to allow access to archaeological sites if you're ever going to successfully build a constituency for preserving them," Davidson said. "But I'm damned if I know how that's ever going to get done. Every time you open a site up for visitors it just brings on all kinds of damage that you really can't accept."

The solution, he believes, may lie in rethinking what we consider proper use of the region's fragile and rapidly disappearing archaeological sites. "I hope we're getting past the idea of telling people that all you have to do with archaeology is visit the sites and go home. Just because you've come and you've been exposed to it and you've taken back some knowledge that you didn't have—that's not enough of a payment. I'm not sure I think that's correct anymore," he said. "We can't put a fence around the area and keep everybody out. But what I do hope is that we get to the point that people come to think that if you visit these resources that you have some kind of inherent responsibility to make some kind of payback pretty directly—where you will come and help document a rock-art site or spend some time picking up trash. You might help out archaeologists working on an excavation, something like that. I think the public has to take on more of those responsibilities. But those kinds of changes in attitude don't happen overnight," he said.

Like many other archaeologists working in the area, Davidson believes that the agencies in charge of managing and protecting these fragile resources are simply too understaffed and underfunded to police and protect them all—and given the area's size and the thousands of sites, that task may well be impossible. "In the end it isn't going to be the government that preserves these sites or the political structure. It's really the people who use it that are going to preserve it," Davidson said. "Of all the resources people come here to see these are the most fragile and the ones most prone to loss. If we aren't willing to take these kinds of actions ourselves, they just aren't going to be here. It's a pretty simple equation."

Leigh Jenkins works for the Cultural Preservation office of the Hopi tribe in northeastern Arizona. Like the public lands administered by the Bureau of Land Management, those of the tribe are victimized by collectors of artifacts and pots—those of ancient ruins as well as religious and ceremonial objects still in use. While his official duties often center around ancient artifacts, privately he often thinks of the present and future as well. While time and vandals slowly carry away the past, contact with the outside world is slowly changing the Hopi way of life as well. As a child in elementary school he can still remember being beaten for speaking Hopi in school. The days of that kind of suppression are thankfully gone, but as Jenkins sees it there are more subtle and serious struggles at work today, a persistent wearing away of culture and custom—even though the Hopi are among the Southwest's most conservative and traditional native peoples. Hopi villages still cluster on the tops of mesas as they have for centuries, but the people themselves are increasingly moving into scattered trailers and simple homes around the mesas' bases. Villages like Walpi on First Mesa have been continuously inhabited for more than eight hundred years, but today only a handful live in the village year-round. While others maintain homes and apartments in the village, they return only during the all-important ceremonies. The scenario is being repeated in village after village at Hopi. While fields of corn, beans, and squash still lie scattered across the desert beneath the mesas, survival today is often no less dependent on the money earned from jobs in Winslow, Holbrook, and Flagstaff, or a position with the tribal government on the reservation itself. The kachinas still return bringing rain and blessings to the people, but the patterns of daily life are changing. Native cultures here are still more alive and traditional than almost anywhere else in the United States outside of Alaska. Even so they are wearing away fast, like bars of sand along the edge of a fast-moving river. Visitors often expect the native people of the region to be either fully modern reflections of themselves or living as their ancestors lived some five hundred years before—as if time had never passed here and the world had remained unchanged. For most, however, life is an often uncomfortable mix of old and new—struggling to adapt to the prevailing ways and means of the world outside while keeping the traditions of the past alive.

In spite of the best efforts of missionaries, teachers, and administrators, the Hopi have managed to keep their religious traditions alive. But while the tribe finds itself increasingly in control of its own destiny—its opinions respected and considered—a sense of being overwhelmed prevails among those few I managed to meet, as if they were refugees on an island watching the tide around them rise. One afternoon in early December as a lid of low gray clouds covered the mesa, dusting the air with flurries of snow, I spent a few hours of an afternoon in Baccavi talking with Jenkins and his friend Riley Balenquah about the persistent pressure of change.

"Western influences have been around for a long time, but it wasn't really until after the end of World War II that the Hopi began to be impacted," Jenkins said. After the Japanese attacks at Pearl Harbor the Hopi like other native peoples were pushed to enlist and drafted into the armed services. For many the war brought not only their first real contact with the world beyond their mesa and the still isolated deserts of northeastern Arizona but a cash economy as well. While the men went off to war, the women were recruited to work in assembly lines and factories. "That was the start of the change for Hopi," Jenkins said. Those who came back from the war and the factories were suddenly accustomed to working salaried jobs with regular paychecks. Training programs organized by the Bureau of Indian Affairs appeared, teaching everything from jewelry making to ship welding. Some of those programs relocated families to distant cities like Dallas and Oakland. With the veteran preference in hiring and the economic boom that followed the war, others found jobs in nearby towns such as Winslow, Holbrook, and Flagstaff with army depots and logging operations. Those who had learned how to tend boilers in ships during the war found work tending the boilers of steam locomotives with the nearby Santa Fe

Railroad. "Eventually people began to move away to live closer to their jobs. Their home life changed," Jenkins said. Balenquah, for his part, can remember going over to visit relatives when they came back home from jobs off the reservation. They would, he recalled, bring back things he had never seen or imagined. "Maybe I could get a taste of a hot dog or something like that."

In the 1950s and 1960s the Hopi began to invest in trucks and tractors for their farms. "All of this built up kind of a momentum shift to cash-based economy," Jenkins said. There is no doubt that as a people they are becoming more and more integrated into the world outside. The desert that surrounds them is spare. While it kept their ancestors alive for centuries, the modern world demands far more than it can possibly provide in the terms of cash and property.

"It's really hard to accept as a Hopi that changes are taking place—and that in some cases we contribute to that change. But while our culture may be struggling, we are blessed with a culture that is still viable," he said. "We're blessed with that, but all of us who are parents today, my age group, it's probably more a challenge to be a Hopi today than it was for my grandparents.

"Let me explain it this way: I can never really feel, much less understand, what it took during my grandfather's time to survive on your own—meaning out of your corn and your flock of sheep. They literally had to struggle to survive. Probably their spirituality demanded a high degree of discipline. Today we're subject to this cash economy. Me and Riley work eight hours a day, five days a week. Every two weeks we get paid for eighty hours of work whether or not we worked for eighty hours or loafed around for a couple of hours. For me it's twice as hard to be a Hopi because not only do we have to deal with this transition or change that's occurring—probably every day—as Hopi people; but we also have to deal with being a member, whether we like it or not, of a more dominant society. We are part of it. We have to survive economically.

"Me and Riley were talking a while ago, saying, 'Boy, I wish we could just quit our jobs, go down to the valley, plant orchards and farm and survive just like that.' Maybe me and Riley are willing to do that. I don't know. But it dawned on us that in order for us to do that we've got to change the attitudes of our kids and spouses, our parents even.

"Our grandparents survived famine and droughts. I can't even go half a day without feeling hungry. My grandfather went for days without food. It's a credit to those people, but I think people today have a dual challenge—to be Hopis and to deal with the real world out there."

Balenquah generally agrees. "It's difficult, but on the other hand I think we have one other option. We're blessed and fortunate to have this still viable culture and identity—even though it may not be an exact performance of whatever function Hopi does. We still have a pretty good idea of how things went.

"Now I have the English language fairly well in command. I can probably struggle through a lifestyle with the dominant society and live off the reservation. We have that option too. Forsake the life at home and go out into the concrete jungle and be like everybody else. At that point, if the person is strong enough to focus on it and devote themselves to it, it's pretty easy to melt into that pot. But the real responsibility of your family, where you come from, really takes a far-back second seat. It's difficult to live in two societies. It's like the song says: One's got my name and the other one's got my heart. You know? The question is, where's your heart at?"

Today they constantly find themselves trying to rethink just where they are. Jenkins recalled the experiences of his father, who was governor of his village in the 1960s. He resisted paving the village roads because he just didn't see a traditional Hopi village with blacktop on it. But the BIA engineer told him, " 'Sorry, Governor. You've got a once-in-a-lifetime opportunity to have your roads paved, but you've got to do it our way.' That's what they told my father and so they blacktopped the whole village with no consideration for drainage or anything like that. So now we've got homes that get flooded every time it rains," he said with a laugh. Those kind of things probably wouldn't happen today. "I think we're reassessing a lot of what's occurred and yet, I think there's probably a little bit of a yearning to try something else other than what we've been forced to live.

"By the time we got out of high school my father always told us you need to have an education to survive—meaning the white man's way. That was so ingrained into his generation. They went to BIA schools and had to dress in military clothes, get their hair cut, salute and march around. And they were told: 'You are going to become white men.' But that was the kind of influence that dominated past generations.

Because they were beaten into submission. He always told us that you've got to get an education—meaning learning the white man's way. Later he admitted that it never did occur to him to tell us that we had to continue to learn to survive the Hopi way as well."

"Leigh's father said the same thing my father did: The only way to compete, the only way to survive in this world, is to get an education," Balenquah said. "Now I rebel. Let's define education. Formal education versus street education. Who's the more successful person?"

Today Balenquah and Jenkins find themselves questioning almost everything that comes their way from the outside world, even well-intentioned warnings about health and safety. Lately the village has been receiving warnings from the BIA that their springs and water supplies are contaminated with bacteria. They find it alarming, but for very different reasons than most.

"When I was a kid I used to drink out of the same water hole as my sheep and never got sick," Balenquah recalled. There was piss and crap in there. "Now I'm sure that if I even drink from a spring out here on the mesa I will get sick. I will make myself sick because I don't have enough faith in myself. I'm too good for that natural spring out there. I question myself at times. Why can't I go back and be the innocent little kid I was and not question anything?"

It's these kind of subtle influences that tear the Hopi away from the past. "We still place offerings at springs," Jenkins said. "And yet I have this mental barrier that I shouldn't drink from you. One of the teachings of our people is that you must never doubt the integrity of something. It is, in fact, a blessing for you. That's one of the teachings of the Hopi: that you appreciate everything. Even one iota of that kind of feeling—suspicion—and you destroy the spiritual essence of that blessing."

Along with their religion, however, the land has been their saving grace. "My own opinion when you really get down to the bottom or essence of Hopi faith, to me, it deals with how much value you place in a harsh and difficult agricultural lifeway. When the Hopis came into the present world we had a choice whether or not we would accept this kind of lifeway. Hopis say that when the clans emerged from the Third World they came upon a spirit or deity, someone they called Masawa, and they, we, recognized that he was in fact the ruler over this domain of the present world. Our traditions say that upon that recognition, when we asked that guardian spirit if we could stay with him, he, Masawa, merely looked at us and said, 'It's up to you.'

"What he offered as his material possession to the clans is something to me that is very significant at how I look at agriculture and farming. When he said your own choice, that it was up to you, he meant if we were willing to live his way of life. The material possessions that he showed the clans were a planting stick, a quart of water, and a pouch of seeds. And he said, 'That is my life and this is my world here. This is my spiritual center.'

"So the Hopi clans that are seen as Hopi society today on the Three Mesas have in fact chosen to accept that kind of lifeway. It depends on a significant amount of spiritual discipline. It requires, really, a final acceptance that mankind, meaning you and me, are really at the mercy of nature. And in order to benefit we need to respect it. This is the reality of what this teaching means to someone like me. So I see farming as a sacred duty. To me a Hopi male cannot fully see himself as a Hopi without farming. Noting all of the other religious kinds of things available to us, to me the real ultimate test of faith is whether or not an individual accepts that agricultural lifeway.

"I've chosen to do that. But in the Hopi way let someone else judge me if I'm a Hopi. That's one basic thing. Never call yourself Hopi. Let others decide if you deserve to be called Hopi. Today I still farm all the land that my father and grandfather farm.

"Everything, at least in my family, is still being done by hand. We plant it, maintain it, weed it, hoe it, and harvest it by hand. Our water comes strictly from the clouds. I have a passion for farming. It's something I enjoy. It fulfills my ego. It makes me feel good and it makes me understand why I have to do the ceremonial part of our culture because the two go hand in hand. It produces a sense of balance. It is a place I can escape to, a place where I can find peace and tranquility and know that the kind of work you're putting into it will be returned to you."

• • •

Behind the store the women are at work. Huge pots of dye bubbling over the fire are filled with skeins of yarn in orange, shades of red, and blue. Others hang on lines outside to dry in the sun. Inside weavers work at their looms, shuttles clacking back and forth as they bring rugs and blankets to life. Out front visitors browse through a gallery and store, eyeing rugs and woolen jackets. The store and weaving studio is part of Tierra Wools, one of several businesses started by the village cooperative Ganados del Valle in Los Ojos, New Mexico, a small Spanish village in the Chama Valley some eighty miles from Taos on the far side of the Tusas and Brazos mountains not far from Tierra Amarilla. The studio and store sit right in the center of town. Fields and pastures are found along the town's edge, stretching off the rolling, pine-covered hills and peaks that lie to either side of the broad, high-mountain valley.

Little more than a decade ago Los Ojos was simply another poor but picturesque Spanish village in northern New Mexico, struggling to keep itself alive. In 1983 a small group of village residents decided to band together and establish a nonprofit cooperative that would raise sheep and employ housewives weaving rugs and cloth from the wool in traditional local patterns. Their first year their total sales were roughly fourteen thousand dollars. By 1994 they were pushing three hundred thousand dollars, selling blankets, clothing, and rugs woven from the cooperative's sheep in traditional Spanish styles and designs. While the weaving cooperative is the group's most visible business, they are also closely tied to their flocks. Slowly but surely they are working to rebuild their flocks with the original churro sheep, brought over from Spain in the late 1500s, whose wool is now highly prized. This is not a New Mexican version of Colonial Williamsburg or a sideshow for tourists but a thriving business that has drawn on traditional industries and crafts of New Mexico's Spanish villages in hopes of preserving the town's integrity and creating a source of local jobs.

The transfer of the Southwest from Mexican to American control at the conclusion of the Mexican-American War in 1848 was as devastating for many of its Spanish villages as it was for their Native American neighbors. Treaty agreements required that the U.S. government respect existing property rights. Villages like Los Ojos and others in New Mexico had been set up by a system of land grants— charters granted by the king or governor to establish a new town or settlement. Grants were large, sometimes amounting to several hundred thousand acres. Settlers, however, held title only to small patches of irrigated fields on the edge of their town or village. Grasslands and forests beyond the reach of their cultivated fields were held as common ground, a village cooperative, or *ejido*, open to all for grazing, firewood collecting, and hunting—a style of settle- ment and living that was, as mentioned before, in essence a close replica of that of their pueblo neighbors. Irrigation ditches, or *ace- quias*, were often built and maintained by the village in common. Herds, too, were run in common. The center of the village, in turn, was the village church. Half a world away from Spain and separated from the richer colonies in Mexico to the south, Spanish settlements in the New World were forced to live independently and off the land, just as their native neighbors were. They grew their own food, made their own clothes, and wove their own fabric. Ultimately they would be forced to manage their own churches and govern them- selves as well.

Their indifferent merger with the United States soon unraveled this carefully constructed fabric of community life. As David Lavendar describes it in his book *The Southwest*, "Fuzzy prece- dents in Spanish law seemed to say that although the king might permit villages to use a designated stretch of land in common, title remained with the sovereign. Thus in acquiring the Southwest from Mexico, the argument ran, the United States had also acquired title to the ejidos and could place them in the public domain where they would be subject to homesteading by any American, native or natu- ralized." In short, while private property rights were considered sacrosanct and inviolable, the land's new leaders refused to recog- nize the property rights of groups—at least those that functioned essentially as communes rather than corporations. *Ejidos*, quite bluntly, were somehow seen as "un-American." Suddenly Hispanic farmers and ranchers had to compete with Anglo cattlemen for the right to graze their flocks and cut wood at cost on lands that they had always treated as their own. Licenses were needed to hunt and fish. When it was all over the Hispanic population of northern New Mexico lost some 3.7 million acres of *ejido* land—much of it to spec- ulators and the politically well connected who built up large cattle empires and land companies. While the Spanish peoples never suf- fered the religious persecution and cultural suppression the Native Americans did, they had their land pulled out from beneath their feet while the Indians were settled on their reservations. Suddenly residents found themselves trying to survive on the small garden plots within their villages. With Anglo cattlemen in Texas and Colorado and Indian reservations in Arizona and New Mexico there was no "new land" for settlement or colonization. As families grew, their plots grew successively smaller and smaller as they were divided among successive generations. By the 1920s and 1930s the counties of northern New Mexico were among the poorest in the United States.

World War II brought some changes. Federal laboratories at Los Alamos provided employment for hundreds—particularly in the villages that clustered between Santa Fe and Taos: Chimayó, Cordova, Truchas, and so on. Others joined the military. Land, however, was still the key to everything. Without land many of the region's villages were beginning to fall apart as residents moved off to Albuquerque, Los Angeles, and Dallas in search of employment.

Anger had exploded in the late 1800s in San Miguel County with the appearance of a Spanish vigilante group, the *Gorras Blancas* (White Caps), who burned barns and killed livestock of new grant owners in the county. Later around 1912 a similar group appeared in the Tierra Amarilla and Los Ojos area called the *Manos Negras* (Black Hands), who used similar tactics when fencing began to seal off the area's once common land. In 1966 protests erupted again in Tierra Amarilla, partly in response to the organizing ability of Rejes Tijerina, an evangelistic preacher from Texas who began leading a fight to reclaim the village's original 594,515-acre land grant. Barn burnings, ranch burnings, and fence cuttings finally culminated in a courthouse raid. The New Mexico National Guard was called out, and Tijerina and his followers fled up into the mountains. Their armed struggle would, in the end, achieve little more than publicity. But although their small uprising would be overlooked and quickly forgotten in much of the United States, it would show the rest of the region just how deep their feelings ran.

Land is still a constant problem for Hispanic villages in northern New Mexico and Los Ojos is no exception. In 1989 when the village cooperative was unable to find either private or public rangeland in the area, they decided to run their sheep on the grounds of three large state wildlife areas in the surrounding mountains—land all part of the disputed common ground of the old Tierra Amarilla Grant. They made no secret of their plans or intentions, but for a few days their mini-revolt captured headlines around the Southwest. But while the move brought attention to their plight, in the end it brought no solution to their desperate need for land. Although it was widely seen as an act of civil disobediance, the moves excited the ire of a number of conservation and wildlife groups—anger that left Ganados members almost incredulous. "How can somebody from Santa Fe or California tell me how to run sheep or take care of the land?" said Joanna Terrazas, a weaver and Ganados member, told me as we talked over cups of coffee in the cooperative's craft shop and bookstore a few doors down from the weaving studio. "We've lived here all our lives. We have to take care of that land. The sheep and the wool are all part of what we do here. The land is what brought people here in the first place." Even more galling is the total ignorance many outside the area show for the region's long Hispanic history. "People stop by and ask us what tribe we're with," she said.

While cooperative members desperately look for rangeland for their flocks, they are also discussing plans, typically, for taking things into their own hands—not with another act of civil disobedience, Terrazas explains, but by setting up their own land bank with a revolving fund so locals and co-op members can afford to buy up privately held rangeland and farmland in the area and gain some control over their future. Land is the key to everything, she said, and they will have to move quickly. "Land around here is selling for twelve thousand dollars an acre," she said. "How can a young guy with a family hope to afford any land?" A proposed ski resort outside of town has added to their problems as well, as subdividers and developers have begun splitting up ranches outside of town with prices that are geared more toward the potentialities of condos and cabins than cows and sheep.

"I know we need change. But it's scary and to families that have roots it hurts a lot." The newcomers, she said, have far more money than locals and are far more possessive and far less likely to share. "You drive around now and you see NO TRESPASSING signs everywhere. It used to be when somebody was local you could go to his land and pick some stones or cut some firewood and nobody cared. Now they don't want you there. "That's my land," they say.

While outsiders tug at the fabric of the community, the cooperative is striving to provide jobs and a sense of possibility to locals—a way of maintaining their ties both to the land and to the traditional crafts and skills that were once the key to survival. So far it seems to be working. "More and more people are staying and it's spinning off into other things. Kids realize that there is a job in agriculture here, that there are jobs in marketing here, that there is a job in business back home. We can get kids to realize that they can do things here. They don't need to move off to Santa Fe and Albuquerque," Terrazas said. For other small villages both here and around the United States, Los Ojos has become something of a blueprint for survival, with calls and letters from others coming in regularly to ask for input and advice for their own plans to pull themselves together. All that is encouraging for Terrazas, but the most important thing for her is her art. "I want to weave," she said. "That's my love. Once you get started you can weave all day. I don't want to stop."

At the end of the dirt road that leads to Jayne Belnap's house in the Spanish Valley south of Moab, we waited for ten minutes for the traffic on Highway 191 to clear—a steady stream of motor homes, jeeps, and cars laden with mountain bikes and camping gear in improbably bright shades of purple and green. "A few years ago I never even had to look when I came to the highway," Belnap said. "It just keeps getting worse every year." Over the past five years Moab has become the Southwest's latest tourist hot spot. Thirty years ago it was a small and almost unknown mining and ranching town in the midst of the slickrock deserts of southeastern Utah. Today it is flooded with more than 3 million visitors a year. Photographs of the brightly colored arches and spires of rock in the deserts outside of town now grace the covers and pages of magazines ranging from *Outside* to *Newsweek*. Abandoned mining roads and once-deserted trails are now filled with scores of bikes and jeeps. Images of Delicate Arch, a solitary curving arc of slickrock in the heart of Arches National Park just outside of town, now cover the sides of billboards and license plates. The scenery here has become a resource to be exploited like oil or gas or open range.

Belnap is a biologist with Canyonlands National Park. Although her work often takes her far afield, she sees signs of the region's sudden surge in popularity almost everywhere she looks, not only in the steady stream of traffic that shoots down the highway, but out in the midst of the desert itself. Belnap's particular area of expertise is cryptobiotic soils—the black sheets of cyanobacteria, lichen, and moss that cover the floor of the desert here, an anchor for soils and a source of nutrients for plants. Over the past ten years as the crowds have grown, these pioneering crusts have been disappearing into thin air, like the curtains of rain from a late-summer thunderstorm that evaporate without ever reaching the ground, trampled underfoot by a steadily growing army of travelers and tourists.

It is late September and I am traveling with Belnap and her assistant to a study area in the deserts outside of town to look at cryptobiotic soils and the signs of damage and abuse. The whole area, she says as she drives, is spiraling downward fast. "The BLM and the Park Service are headed down disaster lane." As we head south down the highway the slickrock desert flashes by outside the windows at sixty miles per hour, a blur of green pinyon-juniper and bright red rock. At high speed, the land beyond the highway looks almost untouched. there are no houses or towns or even fences. A few miles farther south we turn west onto an unmarked dirt road that leads back into the slickrock, an area just outside the boundaries of the park known as Behind the Rocks. A maze of tracks leads off the main road, winding through the desert scrub and back into the rocks, as if a battalion of tanks had recently passed by while out on maneuvers. As we park Belnap's well-worn jeep and begin walking, we weave between scattered low stands of pinyon and juniper and thickets of cliffrose, following pools of pink sand. "Five years ago most of the ground in between these trees was black with cryptogams. The ground here shouldn't be pink, but chocolate brown." After a few minutes walk we away from the edge of the road we finally find what we have been looking for, a healthy patch of cryptobiotic soil, its surface dark and crusty, like a layer of mud or clay left by the evaporation of a muddy, weed-choked pool. Here and there lichens and mosses cover the black soil with tiny patches of color. "A few years ago I used to be able to find healthy soil surfaces just ten minutes from my house. Now I have to drive a half-hour or more and even when I get there I really have to look," she said.

While the trees and shrubs here still look healthy and green, without the crusts of cryptobiotic soil like these beneath them, they could soon be gone as well. While living soils are perhaps the least noticeable part of the landscape here, they are also quite possibly the most important—pioneering plants that anchor the loose desert soil and fix nitrogen out of thin air, the stable base of the desert's flora. Predicting what will happen when these fragile plants disappear completely is an uncertain task, but the implications in Belnap's mind are troubling. While the grasses, shrubs, and trees of the desert here may be able to survive on built up stores of nutrients in the soil for years or even a few decades, the system will almost certainly collapse. While other plants may eventually move in to take their place, the resultant landscape will be far different than it is today and, Belnap believes, far less rich and far less diverse. Overuse is threatening not just to destroy the region's delicate cryptobiotic soils, but to pull the floor out from under the entire ecosystem.

More than a century of grazing has reshaped the face of the desert here. While cropping by cattle and sheep has changed the distribution of native grasses and shrubs, their hooves and feet trampled fragile cryptobiotic soils. Over the past ten years the sudden surge in tourism has taken that damage into new and previously untouched areas. "People are worse than cows as far as cryptobiotic soils are concerned," Belnap said. "Cattle like water and dislike slickrock so most of their damage was limited to stream banks and creeks. People, however, like slickrock and carry their own water and they travel almost everywhere."

While the cryptobiotic soils are able to thrive on the harsh world of the desert floor, the plant-like cells within them such as cyanobacteria or blue-green algae are brittle when dry and easily damaged. A single footprint can kill off several hundred years of growth, compacting the soil and trampling the tiny plants below. Left undisturbed, these living soils can recover, but the process is slow. While thin new crusts of cryptobiotic soil can spread across the surface of a damaged area in the space of a few years or decades, rebuilding the all-important network of roots and filaments that once ran through the ground below can take more than a century. The lichens and mosses that eventually make up these crusts may take even longer to recover—a thousand years or more Belnap believes, although she admits that number is just an optimistic guess because so far no one has seen even the slightest sign of recovery in trampled areas. Not only are the tiny plants and cells inside these cryptobiotic soils the relics of ancient forms of life, the black crusts themselves are old as well, the products, perhaps, of several thousand years of slow and steady growth.

Today those crusts are disappearing at the rate of several thousand acres per year. While walking or riding on established trails causes no new or additional damage, the landscape here is open and invites exploration—fins of sandstone to see, canyons to climb, and welcoming places to camp. As bikes, cars, and people wander across the ground, they leave a web of new paths and trails behind—routes that will soon be followed by others who will spread the damage even further. The problem is occurring not just at places outside the boundaries of parks like Behind the Rocks where Belnap and I have gone today, but inside them as well. "I've got spots in the Fiery Furnace in Arches National Park that I've gone to for years that used to be plant-covered, soil-covered hillsides that are now nothing but rock and sand," Belnap said. "You'd never know it if you hadn't been there before."

Those traveling through the area, however, have a hard time grasping just how far the damage reaches and just how serious the situation is. "The scale of the landscape is so large that people just can't believe they could use it all up. People come here from L.A. or New York or Salt Lake and they go on a trip and they don't see anyone for several days. Compared to what they're used to it's vast. What do you tell them? We're losing this whole area big time, but I don't think people understand that when they come to visit. They see acres of barren sand and they think it looks normal. They think it should be a desert and deserts are dirt and rocks, so it's OK if there's nothing but dirt and rocks," she said. Belnap's job does not center around forming policies or making rules, but in study and evaluation. Although she is passionately concerned about the decay she sees in the landscape around here, she does not envy the task of land managers in the area who strive to balance the needs of the land with those who use it. What she has been able to tell them so far is that for much of the area, its carrying capacity for people is essentially zero. Understaffed and underfunded federal land management agencies in the area can do little more than watch. "It's a moral dilemma really. In some ways it's worse having regulations you can't enforce than none at all," she said. "At the bottom of everything the problem is still people. Using this area is going to damage it. Even me. And I'm careful and watch where I step when I go out, but even so, I have an impact. I put myself in the same boat. If I were in charge of managing the Colorado Plateau I would buy up all the land and throw everyone out. Myself included. People were never meant to live here in large numbers."

I did not make it back to Moab to talk to Jayne Belnap until the following summer. She had moved from her house amid a grove of trees in the Spanish Valley south of town to another up amid the rocks overlooking town, not far from the Slickrock Bike Trail. She no longer visits her test plots in the Behind the Rocks area near her old house. In less than a year the cryptobiotic soils she studies had all but completely vanished. "It's just terrifying," she said "In one year the whole place got driven over—a couple of thousand acres." The culprits, she explained, are not just tourists, but locals as well: everything from woodcutters out for firewood to teenagers with jeeps and dirt bikes in search of fun. With local land management agencies critically short of staff and funds, the area was seldom patrolled. "Aside from the pinyon and juniper, there's nothing out there to stop you. Blackbrush isn't going to stop you. You just run right over it," she said. "I wanted to study things on a landscape level, but I'm finding that the landscape is already gone."

Moving up outside of town with a view that stretches for several miles has given her a new perspective on her work and the changes taking place in the area. "When I was down in the trees I was just duking it out. I couldn't really see. Up here I see the big picture. I see the highway and the cars. Voom. Voom. Voom. And I see the sand blowing and suddenly I can see the magnitude of what I have been thinking of." Although she has been watching the area's deterioration for several years, the sudden speed of things has taken her by surprise. "I see things a lot more clearly than I did. I think too that I'm giving up hope. I didn't have much hope to start with.

But I think I'm giving up what little I had. What does it matter if I help save this place for an extra ten years? It will still be gone. I'm never going to save this place. I might just help have its demise slowed down."

In 1948 an uncle of Belnap's wrote an article called "Is Utah Sahara Bound?" decrying the overgrazing and overuse that was destroying the state's rangeland. While she herself is more than alarmed about the signs of decay she sees around her today, she is reluctant to be a voice of doom. Predictions by other scientists of population bombs and global warming that failed to materialize have given her a healthy skepticism about the speed of ecological disasters. While the region's cryptobiotic soils are disappearing quickly, the worst effects of that loss may well take years to appear. "I don't think we're on immediate doom scales," Belnap said. "We're going to tick down nice and slow and no one is going to notice. They're not even going to notice what's been lost because the last person who knew what the area was supposed to look like died."

That loss of memory is something she thinks about regularly. While more visitors and newcomers flock to the area each year, few have any real understanding of its history or fragility. Land management agencies in the area face the same problem, rotating staffs and shifting personnel from area to area and park to park. "They move you. And they keep trying to get you to move," she said. "But if you don't have a sense of history, how do you know what's changing?"

Meanwhile, Belnap said, the area continues to sell, drawing tourists by the thousands. Beautiful pictures and photographs offer a glimpse of striking scenery, but they fail to show just how much is being lost. "All the photos in all the books on all the coffee tables, what they're telling people is not the truth. They're telling people that it's OK out here and it's not OK out here. The place is being pulverized and lost. I want people to hear that sadness and frustration. I'd like them to know that. We have these pictures of Arches National Park that show Delicate Arch without any people around it. We shouldn't be doing that. What we should do is have these pictures of the arch with 14 people beneath it and another 30 across the bowl of sandstone in front of it—because that's what you find when you go up there. I don't want to say that people don't belong here. People do belong here. But in very small numbers."

Although she grew up in Salt Lake City, Belnap spent her sum-mers in Moab where her father took the family to hunt for uranium. They travelled widely in the canyons and slickrock deserts outside of town "looking," as she termed it, "for our riches." While that rich strike never materialized, it gave her an attachment to the desert; its openness and space. She did her undergraduate work at the University of California at Santa Cruz and her master's degree at Stanford, and then she came back to Utah to finish up her Ph.D. at Brigham Young University, working with Kimball Harper. While she still travels widely, a need for space has always brought her back. Like other locals, however, she has begun to feel that the land-scape is shrinking around her. "My favorite places I can't go to any-more because they're full of people. Or the remains of them. Which makes me even madder. My personal space is just gone."

While Belnap is troubled by the crowds, she understands the forces that pulled them in—both the appeal of scenery and the eagerness of locals, at first, to bring them in. After the uranium boom went bust, the town all but disappeared. "What we had were desperate people," she said. "They were going to lose their homes, they were going to lose their lives, they were going to lose every-thing. It's kind of like starving kids that can't get enough." No possi-bility for jobs or employment was overlooked. For a time local officials even lobbied for putting a nuclear waste dump in the desert south of town. It would, they argued, bring not just jobs, but tourists as well. People would pay to see it and take tours. No pro-posal was too far-fetched or too fatuous to be considered. Uranium made a brief recovery in the 1980s, only to collapse again. In the 1990s tourism in the area, which had been building slowly, finally exploded. Instead of a few thousand tourists each year, the area was suddenly flooded with several million.

For those who lived through the town's hard times, the sudden surge in popularity was a welcome relief, but also one that was filled with problems, straining roads, schools, utilities, and even tempers. Growth in the area has taken place so rapidly that locals had not time to consider what they might be losing; no time to consider what they might be asked to give up; no time to consider what type of community they wished to have; no time to plan. For Belnap no less troubling than what the growing crowds of people have done to the fabric of the landscape, is what those same crowds have done to the fabric of the community. Like the native and Hispanic communities around them, locals here have also begun to feel cramped and crowded by the world outside, a world that seems to come by with increasing frequency and seems increasingly inclined to stay. The problems here are no different than those taking place in other once-remote tourist towns throughout the region: Santa Fe, Durango, Sedona, and Taos. The only difference here in Moab is that the changes have been so recent and come about so quickly.

There is a subtle difference here, however, between the uneasiness brought by crowds and outsiders in Moab and the uneasiness felt by nearby native and Hispanic communities. What seems to trouble locals most is not so much the loss of a particular culture or community, but a loss of space. While the first Mormon pioneers who settled in the deserts here came to join a community, more recent arrivals have often come to avoid one—exiles and refugees who saw the vast open reach of the landscape here as a last chance to avoid the pressures and confines of the more settled world outside. The last good place.

The land here has meant many things to many people: ranchers, rafters, miners, and hikers. In the past, however, there was always enough land to go around, for the conflicting interests of those who come here to work and those who come here to play to avoid each other or make allowances. Over the past decade, however, that sense of space has been shrinking rapidly and the conflicts are often explosive. While the antics of environmental activists from groups like Greenpeace and Earth First! capture headlines by spiking trees and harassing whaling ships, southeastern Utah has its own activists as well—not those who seek to preserve it, but those who push for development and growth and are angered by the restrictions of parkland and wilderness, not Earth First! but an underground and unorganized movement that should perhaps be called Earth Last!: angry locals who deface pictograph panels with sledgehammers and spray paint; and others who use bulldozers to drive roads into remote canyons or level pueblo ruins to protest everything from the designation of wilderness areas to the arrest of prominent area residents for pothunting and stealing artifacts and bones. Other attacks are more personal: phoned-in threats and small-town Fourth of July parades where the featured float behind the high school marching band shows a local park ranger being hung in effigy.

While Belnap has no sympathy for their actions, she can understand the source of the anger and hatred behind it. It has to do with an illusory notion of independence and the myth of rugged individualism that pervades the West—stronger now than it ever was in the days of pioneers. "The one thing I got told when I was growing up is 'Don't ever let anyone tell you what to do. That is your right as a Westerner. To never have anybody tell you what to do.' Now all these people are moving in and telling them what to do. Their whole world is changing and they don't know why."

Recent immigrants arrive expecting to find a quiet oasis, only to discover that the clashes between people of different cultures, colors, and classes here are even stronger than those they left behind. "Everything you see here is what's in everybody," Belnap said. "It's just magnified because they've been able to get away with it for so long. They didn't have the population pressures that forced them to subsume their particular feelings or emotions. But the fact that it's so magnified, I think, gives you a good chance to figure out how to fix the globe right here. It's not hidden under layers of civilization and socialization. You can look at it directly."

In the modern world today we are often as estranged from each other as we are from the land itself. While the spareness of the land amplifies the conflicts and disagreements between those who live here, its spareness also amplifies the conflicts we face with the land itself. Resolving them, however, will not be an easy matter. A good place to begin, Belnap believes, is planning—deciding how we want to use the land and understanding the trade-offs between parks and cattle or mining sites and oil wells; deciding whether in making room for ourselves we are also willing to make room for others—not just other people, but forms of life as well: the grasslands, forests and rivers of not just the desert here, but the entire earth around us. The danger of not doing so is plainly visible here. Whether or not we are willing to do something about it is another matter. Looking out her window at the cars speeding by on the highway below and the hills of loose sand that now blow through the desert, Belnap is not encouraged.

"At some point it's up to us not to have so many of us—or to decide to do with less or to give up some of our individual rights so we can do it better. But we're not willing to do any of that. And realizing that has given me a lot of sadness. That's what this view has given me. That's what I've learned here. Sadness is a good thing though. Sadness is the thing, I've decided, that adds three dimensions to the human character. We're two dimensional if we don't have that. And I've got a lot of that here. Just tons of sadness."

I left Moab that afternoon, driving south on Highway 191 through the Spanish Valley past the La Sal Mountains toward Monticello and the Abajos. Outside of Monticello I turned off the highway and began working my way across the desert on dirt roads and jeep trails, heading southeast toward the Four Corners. Here and there the ground was laced with small canyons and washes, some dry, others concealing fields of corn and beans and groves of cottonwood trees. A thousand years ago Anasazi farmers planted their fields in the same canyons and washes. Today they are home to others: Navajo, Ute, and Mormon. From time to time I passed others, an old woman on the edge of the Navajo reservation herding a flock of sheep, a tank trunk speeding across the desert trailing a quarter-mile long cloud of dust behind it.

Above the narrow reach of the washes and canyons, the ground was wide and open, the grass and dry scrub of the desert a flat floor for the vast sky overhead. Here and there bluffs of sandstone and shale rose up out of the ground in shades of purple and green. Their sides looked almost like weathered copper. Mountains marked the edge of the horizon like compass points, the La Sals to the north, Sleeping Ute Mountain to the south and the Rockies and the Abajos to the east and west.

Parking alongside the road I began walking across the open ground toward a distant line of cliffs until my truck was nothing more than a small point of green in the distance. As the land rose up, thin forests of pinyon and juniper appeared. Coming to the edge of a hidden canyon I climbed down to the floor below and began walking past solitary stands of yucca and the blooms of sunflowers and asters. Rocks that covered the ground above were pale and grey. Those here in the canyon walls below, however, were colored in shades of red and white, their surfaces marked with the fine lines of ancient dunes and the cobbled paths of streams. In the space of a few inches of rock were several thousand years of time. Farther up the canyon, small pueblos appeared, ancient buildings and apartment houses with three and four rooms, some with small towers or a circular pile of rubble suggesting a kiva. Broken chips of pottery covered the ground, colored with patterns of black, white, and red. Up on the rocks above were the images of antelope and deer. Although the canyon here had no name on my charts or maps, several others had been here before. Walking through the soft sand of the canyon floor I followed the footprints of others.

My grandfather attended the University of New Mexico in the early 1920s and spent no small amount of time traveling around the Four Corners area, looking over the deserts and canyons. His brother ended up staying in the area for several years, eventually running the Colorado River from Moab to Needles, long before the arrival of dams and need for making reservations to run the river. Later my grandfather came back to travel through the area with my grandmother as well. Their house was always filled with traces of their time here: Navajo rugs, Pueblo pots, pieces of turquoise, and the bones of dinosaurs as well as boxes of four-by-five inch black and white negatives—photographs of mesas and buttes from which we made contact prints in a makeshift darkroom in the basement. In the 1960s as the waters of Lake Powell continued to rise behind Glen Canyon Dam he used to tell me we should get out there. "There are cliff dwellings and ruins and pictographs all over the place. We should get a boat and some camping gear and go see it. In a few more years it will be completely gone. No one will ever see them again." We never quite made it. At first I was too young and then he was too old. I did not make it out to the Colorado Plateau until two years after my grandfather had died. Glen Canyon was completely gone by that time, covered by the deep blue waters of Lake Powell. I found other places to travel, however, and stayed for nearly two months, traveling across deserts, canyons, and high plateaus. For a period of two and a half weeks I woke up every morning without fail to find a raven circling through the air above my head as I slept in the open, listening to the soft swish of the wind through his feathers in the still, dry air. Up on the canyon walls were herds of bighorn sheep and flocks of pinyon jays. Walking through a nameless side canyon at twilight one night, I remember turning a corner and finding the face of the cliff covered with hundreds of swirling figures—a collection of spirits, shamans, and dancing men. My grandfather would have enjoyed the trip.

That was several years ago. I have since come back to the plateau many times, but each visit also gives me a new feeling of loss—a few more people, a few more signs of others, a little more like every-where else. I have no expectations that the world here will remain untouched, but I would like to imagine that the changes we bring about could be changes for the better. At times I wonder if my own perceptions of the landscape have not been altered by my own changing frame of reference. Since I first came to the Colorado Plateau, travel and work as a writer has taken me to many places, travels to Africa, Alaska, and Mexico that have changed, in some subtle way, my own perceptions of wilderness and openness. Then too I begin to realize that with the passage of time my own recollec-tions have become more like dreams. I remember the warnings of the soft-spoken, all knowing narrator of Harriet Doerr's *Stones for Ibarra*: "Memories are like corks out of bottles. They swell. They no longer fit."

It is far easier to be enthused about the past of the Four Corners region and the surrounding Colorado Plateau than the possibilities for its future. While people have lived here for thousands of years, the conflicting demands being placed upon it today for water, work, and play seem to be pushing the land here to the point of exhaustion. New residents and visitors flock to the area each year, and the region's traditional cultures and peoples are often caught up in the whirl as well. At times the land here seems to be losing both its integrity and its health. The danger here, however, is not so much to the land, but to ourselves. The landscape here has slowly emerged over some 2,000 million years of time, drifted northward through the equator, and survived the onslaught of both ancient seas and glaciers and it will continue long after we have gone. Less than fifty years ago the deserts of the Colorado Plateau were widely regarded as a barren wasteland, a site for reservoirs and radioactive waste dumps. Today it is a treasured landscape of beauty and power, vis-ited and photographed by millions around the world. But while that change in attitude is encouraging, it is also fraught with problems.

While we admire the land's ruggedness, we fail to appreciate its inherent fragility. We are drawn to its openness, but imagine that a horde of several million visitors will have no impact on it. The landscape here is still unsettled, but it is not untraveled—and that is an important distinction.

A brightly colored block of rock, cut by a maze of canyons, it is a landscape unlike any other place on earth. The land is wide and open. In the high, thin air, views reach for fifty or sixty miles—off to the distant lines of plateaus and mesas, unbroken by the signs of cities or towns or even trees. The relationships between things are easier to see here, not only between the land and the plants and animals that cover it, but also between the land and those who live upon it.

That spareness in turn may prove to be the region's saving grace, shocking us into an awareness of the land around us, its striking features capable of opening our eyes not only to the canyons, mesas, and deserts here, but the natural world outside our own backdoors—a simple garden or field, a small forest or an unnamed hill. Most of us today live a life far removed from the natural world around us, conducting our daily affairs inside the walls of homes, apartments, offices, and factories. For all the tools and technology that separate us from those who lived here several centuries before our arrival, however, our own survival is no less dependent on the land around us than that of those who came before us. Rather than traveling through the landscape simply to see new things, perhaps we should strive for a new way of seeing, allowing the history and depth of the landscape to remind us of our own place in things. That seems to be a goal worth striving for. The ancient peoples of the deserts and canyons here not only struggled to survive, they struggled to build a society as well. That ancient past has much to teach us. Here on the Colorado Plateau you can see not only the past but perhaps a way to the future as well.

INDEX

Aa lava, 172–73
Abajo Mountains, 6, 30, 61, 167, 236, 237, 238, 250, 287, 288, 355
Abbey, Edward, 285, 294
Abeyeta, Don Bernardo, 191
Acequias, 343
Acoma, 145, 168, 169, 200, 202, 203
Adams, E. Charles, 162
Agassiz, Louis, 247
Agriculture, 84, 86, 101, 114–15, 128, 135–36, 214, 283
  agribusiness, 291, 320–21
  of Anasazi, *see* Anasazi people, agriculture
  debate over events preceding arrival of, 135
  Hopi and postwar, 338, 341
  irrigation, *see* Irrigation systems
  Navajo, 86, 215, 220, 222, 227, 228, 355
  Powell's ideas on limits of the Southwest to support, 314–15, 316–17, 320
  rain following the plow, idea of, 315–16
  Spanish crops introduced to New World, 211
  transition to, 135–36
Aleut people, 82
Alfalfa, 95, 290
Algae, 13, 54
  blue-green (cynobacteria), 38, 74–77, 347, 348
  oolites, 25
  stromatolites, 15, 54, 74
  terrestrial varieties, first, 74
Alpine world, 261, 267–71

American Indians of the Southwest (Dutton), 160
Amerind peoples, 81, 217
Amphibians of Paleozoic era, 15
Anasazi Heritage Center, 126
Anasazi people, 5, 21, 42–44, 55, 84, 85, 95, 100, 123–58, 171
  abandonment of Four Corners region, 43–44, 131, 145–46, 149–52, 158, 161, 168
  agriculture, 8, 43, 44, 95, 96–97, 115, 129, 131, 139, 142, 355
    methods of, 142–43
  architecture of, 42–43, 129–30, 131, 154
    cliff dwellings, 28, 43, 97–99, 124, 131, 132
  beginnings of, 128–29
  graves and bodies, 152–54
  hunting and gathering by, 143, 186
  irrigations systems, 43, 130, 131, 133, 142
  kivas, 43, 129, 130, 132, 133, 138, 146, 147
  methods of, 142–43
  Mexican civilizations and, 126, 127, 137
  name origin, 217
  pinyon-juniper forests and, 186, 188
  pottery of, 43, 96, 129, 130, 131, 133
  pueblos of, 129–32, 138, 144, 146–49, 152–56
    interrelationship among, 143–44

Mormons, 9, 45–46, 59–60, 117,
    238–41, 272–84, 290–91,
    304, 316, 327, 355
  arrival in Utah, 21, 238–39, 273,
    275–76, 290
  irrigation canals, 45, 239, 304
  Mountain Meadows Massacre, 278
  native peoples and, 90, 91, 240,
    276, 278–79
  Powell expedition and, 302–303
  restoring damage to the land,
    281–82
Morrison Formation, 26, 57–58, 292
Mountain forests, 258–68
Mountain mahogany, 182, 262
Mountain Meadows Massacre, 278
Mount Taylor, 165, 166, 168, 169,
    170, 238, 261
Mueggler, Walt, 263–64
Music Temple, 301

Na-dene people, 81, 217
Narabona, Antonio, 221–22
National Biological Survey, 312, 313
National Irrigation Survey, 318
National parks, 46, 47
  see also names of individual parks
National Park Service, 149, 347
Native Americans, see names of indi-
    vidual peoples
Navajo, 41, 44, 81, 99, 152, 154,
    168, 169, 206, 215–31, 290
  agriculture, 86, 215, 220, 222,
    227, 228, 355
  Americans and, 223–27, 229
  Anasazi and, 217, 220
  arrival in Four Corners region,
    219–20
  arts of, 42, 216, 230
  creation myth, 218–19, 220
  in desert grasslands and scrub-
    lands, 113–14
  exiled, 225–27, 228
  Mormons and, 240
  origins of, 217–19
  pastoral lifestyle, 44, 221, 222,

    227–30, 355
  pinyon-juniper forest and, 186,
    188
  population, current, 229
  pueblo peoples and, 86, 220, 221,
    224–25, 226, 229
  reservations, 26, 215, 227, 229,
    230, 290
  the Spanish and, 202, 215, 220,
    221–22, 229–30
  warfare with, 203, 212
Navajo Mountain, 6, 237, 250
Navajos, The (Underhill), 224
Navajo Sandstone, 24–25, 26,
    33–34, 51, 59, 67, 103,
    166, 174, 243, 313
Navárez, Pánfilo de, 194
Neanderthal man, 82
Nelson, Nils, 134
Nevada, 214–15, 223
Newlands Reclamation Act, 319
New Mexico, 8, 46, 222
  Hispanics' dispute over loss of
    grazing land in, 342–46
  Spanish influence in, 189–92,
    200–215
Nitrogen, 76, 77, 348

O'Keefe, George, 174
Oklahoma, 215
Omaha people, 86
Oñate, Don Juan, 199–201, 208
Opata Indians, 222
Origin and Development of the
    Pueblo Kachina Cult, The
    (Adams), 162
Ouachita orogeny, 66–67
Ouray, Chief, 91, 92
Ouray National Wildlife Refuge,
    324–29
Ouray reservation, 55
Outlaws, 291

Pahoehoe lava, 172–73
Painted Desert, 100, 110, 122, 162,
    236, 308